We have the pleasure of sul

Science's Strangest]
$ 13.00 pb
Tom Quinn

U.S. Publication date October 15, 2005
A clipping of your review
would be appreciated
800-423-4525

Chrysalis
Distributed by
Trafalgar Square
No. Pomfret, Vermont 05053
www.trafalgarsquarebooks.com

Other titles in this series

Boxing's Strangest Fights
Bridge's Strangest Hands
Cinema's Strangest Moments
Cricket's Strangest Matches
Fishing's Strangest Days
Football's Strangest Matches
Golf's Strangest Rounds
Horse-Racing's Strangest Races
The Law's Strangest Cases
Medicine's Strangest Cases
Motor-racing's Strangest Races
The Olympics' Strangest Moments
Politics' Strangest Characters
Railway's Strangest Journeys
Royalty's Strangest Characters
Rugby's Strangest Matches
Sailing's Strangest Moments
Shooting's Strangest Days
Tennis's Strangest Matches

SCIENCE'S STRANGEST INVENTIONS

Extraordinary but true stories from over 200 years of science's inventive history

Tom Quinn

ROBSON BOOKS

For my favourite inventors
Katy, James and Alex

This edition first published in Great Britain in 2005 by Robson Books,
The Chrysalis Building, Bramley Road, London W10 6SP

An imprint of Chrysalis Books Group plc

British Library Cataloguing in Publication Data
A catalogue record for this title is available from the British Library.

ISBN 1 86105 826 8

Typeset bySX Composing DTP, Rayleigh, Essex
Printed by Creative Print and Design, Ebbw Vale, Wales

Contents

Acknowledgements

Thanks to the staff at the British Library, the Science Museum, the London Library and the Stoke Newington Particle Physics Discussion Group, ably chaired by the delightful Gabrielle Koonin. And for putting up with endless emails and misquotations, I'd like to thank Jane Donovan and Sarah Barlow at Robson Books.

Introduction

The mad scientist is one of the most enduring images of the modern world. Take Einstein, for example. We think of him as a genius but also as an eccentric, someone so lost in higher thought that he is completely out of touch with the day-to-day lives of ordinary mortals. In fact, nothing could be further from the truth. Without Einstein's insights into matter and energy – insights that originally seemed completely bizarre even to other scientists – there would be no television, no satellites, no radar, and little of the vast range of hi-tech medical equipment that has helped double the average length of time most of us can now expect to live.

But Einstein is only the most famous of a long line of scientists and inventors, amateur and professional, who have devoted their lives to trying to understand more about the way the world works. Of course, what makes inventing fun is that it attracts both the serious scientist and the wacky amateur – the latter occasionally making a far more remarkable discovery than his or her professional colleagues. The history of science is littered with great inventions that have led to enormous changes in the way we live, but what makes that history truly interesting is the vast range of also-rans – the inventions that never quite made it but were patented by their discoverers in the belief, and hope, that they would make a fortune, and/or change the world. From gigantic crossbows to edible record players, from head-bump measurers and bird-powered balloons to illuminated lavatories, the story of invention is as rich and varied as it is entertaining.

And when you've finished chuckling at the hat gun, the flying submarine and the automatic baby patter, remember that people laughed when Trevor Bayliss came up with his idea for a clockwork radio – for this apparently bizarre notion proved to be a godsend for tens of thousands of poor people around the world. For the truth is that, in the world of strange inventions, what appears to be mad one day, can seem perfectly sane the next. Enjoy!

Tom Quinn

GIANT CROSSBOW

ITALY, 1486

Most people think of Leonardo da Vinci (1452–1519) as one of the greatest artists who ever lived. Certainly his *Mona Lisa*, now in the Louvre, is one of the most famous pictures in the world, but painting was actually only a small part of Leonardo's vast output. His dozens of volumes of notes and sketches – many of which are still unaccounted for – were at least as often devoted to inventions, military fortifications and weapons as to portraits and landscapes.

Leonardo drew plausible parachutes, a convincing helicopter and numerous plans for flood defences. He was specifically employed to put his mind to many practical problems that required practical inventive solutions. But even Leonardo was prone to eccentric visual musings and one of his most off-beat inventions was a gigantic crossbow. Drawings for the crossbow survive and the masses of notes Leonardo made in his notebooks about the project reveal that for some time at least he really thought it might be made to work.

Building a giant crossbow – perhaps as much as forty feet long and thirty across – is not simply a matter of increasing the size of a normal crossbow. The stresses and strains are totally different and Leonardo realised this: his notes for the crossbow reveal ingenious methods for creating a laminated structure that would both bend and allow lateral movement – vital given the distance the projectile would have to be dragged back along the body of the crossbow before being released.

Leonardo's crossbow was not intended to fire arrows but rather large boulders or cannonballs; since much of late medieval warfare in Italy was based on city states attacking each other, the winner in any conflict was likely to be the side that could most effectively breach the enemy's defences. If Leonardo's gigantic crossbow had been made to work it would have hurled boulders at a city's walls with enough power to break through them.

An elaborately geared ratchet and pinion system was to be employed to load or draw back the bow and it was probably at this stage that Leonardo began to have misgivings about the practicality of actually using the device. Given the dimensions, the force necessary to get the bow ready to fire would have been enormous and possibly beyond the capabilities of the materials available then to build such a weapon. Yet Leonardo's crossbow is still a strange and wonderful idea and one that, with modern materials, could almost certainly be built today.

THE MIDAS TOUCH

ENGLAND, 1580

Medieval and Renaissance scientists – or philosophers, as they were then known – frequently devoted their lives to the search for the philosopher's stone, which was said to be the key to all knowledge. It would also give up the power to transform base metal into gold. Efforts to discover the philosopher's stone make modern attempts at man-powered flying machines, water-driven buses and celestial time machines look almost sane.

The celebrated Dr John Dee (1527–1608), Elizabeth I's favourite scientist and man of mystery, is a case in point. There were several attempts to impeach the old man – he seems always to have been old! – on the grounds that his secret investigations put him in too regular contact with the devil and all his minions. But Dr Dee survived the numerous attacks on him and his advice was much sought on matters to do with divination, prophesy and the dark arts. Despite his reputation as a man of enormous wisdom, Dee spent his life looking for the secret of the philosopher's stone and all to no avail – hardly surprising when one considers that a lot of his research involved boiling slugs mixed with virgins' urine and mandrake root.

Hundreds of other early scientists took the philosopher's stone route and at least one anonymous inventor made a 'device for detecting the stone'. Hardly anything is known about the device but it was said to be a cross between a medieval astrolabe (a curious brass instrument for measuring the relative distances of the stars) and an orrery (a clockwork model of the solar system). The unnamed inventor, who would certainly have known of Dr Dee, worked in London and was clearly a better businessman than scientist: his device could never be proved useless since someone would have to find the philosopher's stone in order to prove that the detector was unable to detect it. The inventor could indefinitely claim that his detector had not yet worked only because no one had yet got close enough to the stone for the detector to pinpoint it. It seems not to have bothered anyone that the inventor was trying to sell a device that – if it really worked – he would be far more likely to keep a secret as it would enable him alone to find the stone. It all sounds rather like those modern newspaper advertisements that offer to tell the gullible the secret of making money by writing – if the advertisers really knew how to do it they would write that blockbuster themselves.

Even Dr Dee might have been taken in by the device and, who knows, perhaps he set off down the Strand with it in hand hoping against hope that it would point the way to eternal wealth and fame everlasting.

CLOCKS ON FIRE

ENGLAND, 1690

It's easy to forget that until the second half of the seventeenth century accurate clocks simply did not exist anywhere in the world. Large church clocks, which had existed since late medieval times in Europe, were reasonably accurate and for the illiterate the fact that they tolled the hours in villages, towns and cities was a help in getting to work on time.

It was the invention of the pendulum in the mid-seventeenth century that enabled smaller clocks of far greater accuracy to be built. For the science of horology the invention of the pendulum is of immense significance. It led to a vast increase in the numbers of smaller domestic clocks being made. Table and mantel clocks were built in huge numbers along with longcase clocks of varying degrees of complexity. To get round the problem of telling the time at night, various ingenious methods were used, including a repeating mechanism. If you were in bed and woke up wondering what time it was you could pull a cord that ran from your bed down through the floor to your longcase clock; the cord operated a mechanism which repeated the last hour (or quarter, plus the hour, if it was a quarter-striking clock) so you knew roughly what time it was.

But a few clock scientists came up with an even more unusual – if completely mad – idea. This was to build a night clock illuminated by candles. We know that many of the great early makers produced night clocks but for reasons that will become obvious only two of these clocks are known to survive today. That said, they are quite remarkable. The clock dial consists of a metal disc with the Roman numerals cut out of it. A candle is placed inside the wooden hood of the clock behind the metal dial. The whole dial is turned by the clock mechanism so that each cut-out hour number appears in front of the candle at the appropriate time. If you came into your house late at night or woke in the early hours and looked at your clock you might see the cut-out of, say, the two with the candlelight shining through it and you would know that it was 2 a.m.

The difficulty with these clocks – and the reason why so few survive – is that they were an extraordinary fire risk. If the candle fell over during the night your clock and house would soon be on fire. The night longcase clock does, however, have one virtue – it shows that there is no end to the ingenuity of the inventor; he or she may sometimes fail to see the wood for the trees (as in the case of the night clock) but the effort to create something new and exciting has to be admired.

THE ARTIFICIAL HORSE

ENGLAND, 1750

We tend to think of science as a comparatively modern discipline. Most inventions that seem glamorous or seriously life-enhancing to us are products of the twentieth century or at the earliest the nineteenth, but scientists were busily coming up with all sorts of contraptions in the ancient world, the Dark Ages and the medieval period; the difficulty is that so few inventions survive, relatively speaking, from these periods that it is hard for us to appreciate the full extent of ancient inventors' activities. Yet if ancient science produced little earthenware lamps that were useful or amphora for wine and oil there were probably just as many wacky inventions that never got beyond the prototype stage.

By the late Tudor period a few of what we would now consider odd inventions were being made and used and some of these have come down to us. Several early country houses in Britain, for example, still have what were known as exercise chairs. These strange seats were beautifully made by some of Britain's most famous furniture makers, including Chippendale, and they reveal something else about the distant past that is surprisingly modern. In the same way that we worry about getting enough exercise, so too did our ancestors. It's not true to say that they all sat around getting fat because fat was equated with wealth. A good figure was just as important in the eighteenth and earlier centuries as it is now.

In summer, exercise wasn't much of a problem because most adult males rode everywhere and horse riding is very good exercise indeed, but in winter, travel of any kind could be difficult, particularly in the countryside, so those early scientists came up with the spring-loaded exercise chair.

This is a padded chair fitted on top of several stout springs. The springs run down to the ground where they're attached to a strong wooden base. The idea was that the seat simulated horse riding for those kept indoors for long periods by bad weather.

The rider sat in the chair and rocked himself violently back and forward or up and down as if trotting briskly on a horse. A most peculiar sight, the exercise chair probably worked in a limited way by using up excess calories, although alternately sitting and standing from a perfectly ordinary chair would doubtless have done just as well.

LIGHTNING ROD

ENGLAND, 1752

Most people associate the name Benjamin Franklin (1706–90) with the Declaration of American Independence – he was one of the signatories – but what is less well known is that he was a very able inventor and something of an eccentric, who lived in London for many years. He once, for example, swam on his back in the Thames while paring his nails and wearing a set of wooden false teeth he'd designed himself.

Growing up in Boston, Massachusetts, Franklin found storms, particularly violent electrical ones, endlessly fascinating – so much so that he would chase them across the country in order to ensure more time for his observations. By the mid-1740s he was making electrical machines and had turned part of his house into a laboratory.

By 1747 he was sharing the knowledge derived from his experiments with electricity with a friend in London. He sent his ideas to Peter Collinson who hoped to publish them. Franklin is believed to have been the first scientist to use the words positive and negative in relation to electrical forces and by 1749 he had described what we would now call a battery in a letter to his friend.

Franklin also set up an experiment based on a sort of primitive Van de Graaff generator, a metal sphere that could be rubbed to produce an electrical charge. An iron needle was able to conduct the charge away from the sphere. Franklin became convinced that electricity and lightning were really one and the same thing. He also turned his mind to a method of preventing the considerable damage and loss of life caused by lightning each year – buildings destroyed when hit by lightning, their occupants killed by the huge charge involved.

Thus was the lightning rod conceived. It started as an iron rod that Franklin recommended should be about ten feet long and sharpened to a point at one end rather like a long spear. He was convinced that the electricity in lightning would be attracted to the rod and drawn into it and in this he was absolutely right. To try out the idea he built an extremely dangerous electrical kite and flew it high in the sky the next time he saw a violent electrical storm. He ran across the fields towards the very heart of the storm; he had protected himself from the risk of being killed by lightning by using silk string to attach himself to the kite. The kite itself had a metal key. When Franklin flew his kite he watched amazed as the lightning was indeed drawn to the key but, even with the silk thread, Franklin was lucky to survive the experiment. The charge involved in a lightning strike is so huge that it might easily have travelled down the silk and killed him outright. But he lived and wrote up an experiment that had proved that electricity and lightning were indeed the same thing. Franklin was certain that just as his small iron needle had drawn the charge from the sphere, so a metal spike built against a church steeple or other building would safely conduct a lightning strike harmlessly down into the ground by the side of the building.

The next part of the saga reveals the oddities of scientists generally: Franklin insisted that sharpened lightning rods were best, while his English friends and fellow scientists were equally adamant that blunt-ended rods would be far more effective. Franklin built pointed lightning rods on American buildings while English churches had blunt-ended rods; King George insisted on blunt-ended rods and tried to impose them on America. The Americans' use of the sharp-ended rod was seen as a further act of disobedience!

Today, of course, what was originally seen as a bizarre idea with no practical application is used to protect buildings all over the world.

THE HEAD-BUMP MEASURER

ENGLAND, 1764

James Lock & Co the hatters started making hats in London in the seventeenth century. Since 1764 they've been in the shop they currently occupy at the bottom of St James's Street almost next to the Tudor palace of St James. The shop's interior and its fixtures have changed little over the centuries; – creaking timber shelves hold hats of all kinds and the shop still uses an extraordinary invention – a conformator – to measure each client's head. The details of each head, including distinguishing lumps and bumps, are then kept on file so that new hats can be made to order even if the customer is on the other side of the world.

The conformator is rather like a cage that fits snugly over the head. It has a rim made up of hundreds of tiny bars rather like miniature piano keys. These can move back and forth a short distance according to the pressure exerted by various bumps on the head, which means that when the conformator is removed from the head it retains an exact impression of the shape of the head. This impression is then transferred to paper and the paper record kept, usually for the lifetime of the individual, so that the hat buyer can simply order a new hat without visiting the shop for a fitting each time.

Lock & Co have made hats for everyone from Nelson to Charlie Chaplin. Most famously they invented the bowler hat, which was, until the 1960s, the universal headgear of male office workers in the City of London. The bowler hat actually started life as a gamekeeper's hat, designed for the immensely wealthy Lord Coke of Norfolk, whose gamekeepers were occasionally attacked by poachers; the bowler was, it seems, an early form of crash helmet! How it made the transition to the Square Mile remains a mystery.

Locks' reputation has spread far and wide – they have even been mentioned in poetry: in John Betjeman's autobiographical poem *Summoned by Bells*, for example.

GHOST-WRITING LUNATIC

ENGLAND, 1780

Members of the Lunar Society, which existed in the English Midlands at the end of the eighteenth century, were known as lunatics – not because they were mad, but because they were members of a club that met when there was a full moon (lunar means 'of the moon'). For in the days when all roads were completely unlit they needed the full moon in order to reach the remote house by a crossroads where the club held its meetings.

The Lunar Society was remarkable in many ways: its small membership included some of the greatest inventive minds of the eighteenth century – men such as James Watt (1736–1819), who invented the first truly effective steam pumping engine, the remarkable medical doctor Erasmus Darwin (1731–1802), who was the grandfather of the great Charles Darwin, Josiah Wedgwood (1730–95), the famous and at the time highly innovative potter, and the wealthy industrialist Matthew Boulton (1728–1809).

It was Wedgwood who expressed the central philosophy of the club when he said that they were 'living in an age of miracles when anything can be achieved'. This belief in the power of rational thought and experiment led to some wonderful and some wacky discoveries and inventions, but less well known was the group's discovery of a means to fill liquids with gas, a discovery that gave us carbonated drinks.

One or two of the group's inventions were rather less effective – perhaps the strangest example was an elaborate copying device. Boulton, Darwin and the others had long pondered the problem of being able to write only one book or letter at a time. In the days before carbon paper and photocopies this could be a real headache; the only solution where many copies were needed was to employ teams of clerks. Darwin thought he could get round the problem with an elaborate contraption of hinged rods. At one end the rods were secured to the writer's pen; at the other they were attached to another pen held in a frame. Each movement of the writer's hand across the paper was duplicated by the other pen; after months of experimentation the copying pen was able to make a very reasonable version of the letter being written.

No example of this original piece of equipment survives but we have it on good authority that it worked quite well; the main difficulty appears to have been the time involved in setting up the elaborate device and in keeping both quills inked. The time taken seems to have outweighed the benefit – or nearly so – and Darwin and the others would quickly have realised that so much effort to make just one copy wasn't really worth it and they quickly went back to employing clerks or making their own extra copies.

THE OMINOUS OMNIBUS

ENGLAND, 1829

It is difficult to appreciate now, but the invention and introduction of the London bus was seen as outlandish and even revolutionary when George Shillibeer (1797–1866), copying a service already available in Paris, first began to run his public conveyance between Paddington and the Bank of England.

It all started on 4 July 1829. Shillibeer and his partner John Cavill had made small carriages for wealthy Londoners for years, but hearing about the French carriage that simply gave people rides over a set distance in exchange for money, they set to work to build a carriage that would carry upwards of twenty people – a thing unheard of in Georgian London. Within days of the first omnibus leaving Paddington, the success of the service was assured; it was hugely popular and not just with passengers – crowds gathered along what is now the Marylebone Road to watch the progress of this extraordinary new contraption. It was denounced by churchmen, who were concerned that it would give poorer people, who could not afford their own carriage, ideas above their station.

Shillibeer's omnibus was pulled by three fine horses and could travel at up to six or seven miles per hour – which is the average speed across congested London today.

What astonished those first bus passengers was that they could stop the bus anywhere along its route just by raising their hands. The service was such a success that within a few years other bus companies were competing with Shillibeer and undercutting his prices. Shillibeer eventually went bankrupt and spent some time in a debtor's prison. When he was released he converted his remaining buses into funeral carriages, but buses – now run by others – were here to stay.

Designed as a fairly standard if rather large coach, the Shillibeer omnibus did have some unusual features. For example, the driver had reins fitted to his back which ran through into the carriage. Any passenger wishing to get off simply pulled on the reins to alert the driver to stop.

Among the strange-sounding rules pasted up on the first buses were the following:

> Do not get into a snug corner yourself and then open the windows to admit a North-wester upon the neck of your neighbour. Do not spit on the straw. You are not in a hogsty but in an omnibus travelling in a country which boasts of its refinement. Refrain from affectation and conceited airs. Remember that you are riding a distance for sixpence which, if made in a hackney coach, would cost you as many shillings; and that should your pride elevate you above plebeian accommodations, your purse should enable you to command aristocratic indulgences.

By the time bicycles began to arrive on the scene, horse-drawn buses were providing services all over London and there were several attempts to dislodge the horse-drawn bus and replace it with pedal-driven vehicles.

One anonymous inventor came up with a design – almost certainly never built – that involved a vast, elaborate array of pedals and chains running under the bus. Each passenger's seat was fitted with a set of pedals he or she was expected to turn on sitting down. The patent calculations suggested that even if the bus were only one third full it would still be relatively easy for each of the passengers to keep it going. In his wild enthusiasm for the idea of a pedal-driven bus, the inventor clearly forgot that there would be a major problem if only one or two passengers boarded the vehicle.

MAN'S BEST FRIEND

AMERICA, 1830

The first locomotive engine ever built in the USA for general service was the *Best Friend of Charleston* designed primarily by Horatio Allen with E L Miller of the West Point Foundry for the South Carolina Railroad in 1830. Miller, an enormously talented if idiosyncratic inventor, faced strong opposition to his plan to start building locomotives; unable to raise any money from investors or banks, he used his own money and his own initiative.

Railways were central to nineteenth-century industrial development but in addition to practical inventions they also threw up a whole host of bizarre creations of varying degrees of usefulness. The *Best Friend* was a practical working engine but looked most peculiar: it had a vertical boiler with no fire tubes and resembled a giant beer bottle, but all those who saw it were astonished by this new and marvellous machine. At a time when most people still travelled by horse and had never witnessed anything mechanical that was capable of travelling at speed, *Best Friend* was an extraordinary sight – but not, alas, for long.

The fact that machines were so little understood at this time led to many mishaps and occasionally a serious accident. The fate of the *Best Friend* is a case in point. The only man who really knew how it worked and who understood its limitations was the man who built it and the explanations he gave his staff about how it worked and the intricacies of its operation must have largely fallen on deaf ears. They viewed *Best Friend* – and the reaction was the same for very early locomotives wherever and whenever they were introduced across the world – as a kind of mechanical horse. If you mistreated a horse it might kick you or, ultimately, die, but that was about the limit of possibilities. Machines – particularly iron horses – must be the same. But of course they weren't.

Thus it happened that an inexperienced fireman on *Best Friend* was having his lunch one fine day after a long morning's stoking when his peaceful half-hour was disturbed by the noise of steam escaping from the boiler. At first he took no notice, but then the noise began to annoy him. He put up with it for a while longer and then decided it was ruining his appetite and something had to be done. But what could he do? All he'd been trained to do was stoke the firebox. Then horse mentality entered the equation. If something is making a noise that you don't like, you shut its mouth – and that's exactly what he tried to do with *Best Friend.*

The noise was coming from the safety valve so the fireman sat on it and happily continued eating. Some time later there was an explosion so loud that it was heard halfway across the state. *Best Friend* was reduced to a few tangled pieces of metal and very little of the fireman was ever found. Luckily, however, few other people were around at the time of the accident – anyone within fifty yards of the explosion would almost certainly have been killed.

ALL DONE WITH MIRRORS

ENGLAND, 1831

Scientists employed by manufacturers are often kept constantly under pressure to come up with new ideas to generate income for their employers or at least to devise what might best be described as variations on a theme. It's easy for us to laugh at earlier generations who fell for nose squashers or ear flatteners, but we are just as gullible in our own way – we all had perfectly good kettles but as soon as a scientist and inventor (who shall remain nameless) came up with the idea of a jug kettle we chucked away our old kettles and bought new ones that perform precisely the same function as the old ones.

A superb – if wonderfully batty – Victorian invention was a pair of spectacles that enabled the wearer to lie in bed or on the sofa and read without lifting his or her head an inch.

The spectacles were pretty ordinary in almost every respect but they were fitted with a clever mirror arrangement that meant you could effectively see round corners. You put the glasses on, lay down with your head completely flat (rather than propped up on a pillow or cushion) and then held your book in the normal position – light from the pages of the book was reflected through a right angle by the mirrors and down on to your eyes. The only problem was that if you suddenly had to jump up, the glasses would only show you a view down to your feet. Wearers also complained that they felt curiously disorientated while reading, as if the pages of their books were being beamed to them from some distant planet.

WIRE TAPPING

ENGLAND, 1831

One of the greatest discoveries of all time was also – at least then – seen as one of the strangest. For centuries man had been aware of various forces – the strength of a man pushing a cart, the wind that filled a sail – but they were seen as things with no obvious connection to each other. The same was true of electricity and magnetism: they were viewed as entirely separate phenomena until a poorly educated but brilliant Englishman made one of the greatest discoveries of all time. Michael Faraday (1791–1867) left school knowing only how to read and write and had never been to university. At the age of fourteen he was apprenticed to a bookbinder, a highly skilled trade but one that did little to stimulate what was obviously a remarkable brain.

By chance, Faraday attended a lecture in London given by one of the greatest scientists of the age, Humphry Davy. Davy – whose lamp transformed safety procedures in Britain's coal mines – was lionised in Britain and on the Continent and the young Faraday was fascinated by his lecture. Having listened to the great man, Faraday went home and drew some of the splendid things Davy had described. He then made and bound them into a beautiful book, which he sent to the great scientist asking at the same time if Davy would consider taking Faraday on as an assistant. Davy was intrigued and soon he and Faraday were working together.

Faraday was fascinated by electricity and magnets and while helping Davy he set up an experiment that involved placing a thin copper wire near a magnet and then applying electricity. Faraday was astonished to see the copper wire begin to turn and it kept turning until the power supply was disconnected. Faraday published his results and demonstrated how the wire could be made to move using a completely invisible force.

At the time this was seen as incredible and it made Faraday famous almost overnight. One might have thought that Davy would have been delighted at his pupil's success; in fact, he was jealous that the limelight had switched to his protégé and he let it be known that Faraday had actually stolen the idea from him after overhearing a conversation about it. This was completely untrue but Faraday was intensely loyal to the man who had given him his chance and he said nothing, continuing to work quietly in the background until Davy's death a decade later. Faraday lived into old age and was fêted throughout the scientific world for a discovery that turned out to be the basis for every electrical engine ever made.

ROPE TRICK

ENGLAND, 1835

Railway owners were always concerned for the safety and comfort of their passengers if for no other reason than that unhappy passengers would mean a serious dent in the company's profits. The concern was not always extended to paid staff, however, since they already represented a drain on the company profits. They were viewed as a necessary evil, which may explain a remarkable railway invention of the mid-1830s.

This was at a time when accidents were common simply because railways were in their infancy and it was only when a particular kind of accident had happened that the company began looking for ways to prevent it occurring again. With no signalling, for example, trains tended to collide more often than was comfortable. One railway company decided that the best way to deal with the problem was to send their trains out with the engine well ahead and the carriages well behind.

In 1835 they launched a scheme to achieve this. They attached the engine to the rest of the train by a rope nearly half a mile long; this meant that if there was a collision the engine driver and firemen might be injured or killed, but the rest of the train would have half a mile after the accident in which to slow down with or without (depending on whether he'd noticed anything amiss!) the help of the guard in his brake van at the back of the train. The experiment was eventually abandoned because the rope proved more of a hindrance, particularly as the train attempted to negotiate curves in the track.

SUBMERSIBLE

AMERICA, 1836

Submarines seem so modern that it is difficult to believe that they have a long history, but at least as far back as the eighteenth century attempts were made to build an airtight vessel that could at least be lowered safely beneath the waves. Although metalworking wasn't really up to the task at the time, the basic ideas being deployed by those early marine engineers were pretty sound.

By the mid-nineteenth century the rapid technological developments of the ongoing Industrial Revolution meant a submarine that could go under water and also manoeuvre itself independently of any mother ship on the surface was a real possibility.

One of the earliest and maddest of these submarines was a shallow water submersible invented in the 1830s by Mr Lodner D Phillips from Indiana. Cigar-shaped, with a glass bubble rising midway along the main body, the Phillips submarine was entirely powered by pedals. In fact, the internal arrangement was rather like a recumbent bicycle. The submariner (there was room only for one) sat in the middle of the submarine with his head in the glass bubble while the pedals drove a small propeller at the back through a complex series of chains, cogs and rods.

In clear water with no obstructions the Phillips submarine was capable of a few knots an hour, but its buoyancy tanks were incredibly inefficient (it was constantly becoming upended or nose-diving!). With no lights or radar it quickly foundered in murky water, but it was a step in the right direction.

Phillips – who, incidentally, was a shoemaker by profession – built a second submarine that was hand-cranked along at speeds up to four knots; it could also dive to depths of about a hundred feet but again without lights or navigation equipment it constantly crashed into underwater obstacles. Nothing daunted, Mr Phillips offered to sell the patent for his idea to the United States military. They dismissed the submarine as an idea with no future and explained that the United States Navy was in the business of floating on, rather than under, the sea.

GUNNING FOR POACHERS

ENGLAND, 1837

Shooting, like hunting and fishing, has always produced – or attracted – oddballs and inventors and none more so than Colonel George Hangar, who ran his own enormous shoot, but never invited anyone to shoot it with him. After several poachers were chased off his property and his breeding pens were disturbed he decided to take matters into his own hands. He wrote to a friend explaining that he intended to invent the ultimate anti-poaching device, based on his military experience and absolute belief that defence is the best form of attack.

First he had a six-pound cannon specially made, which he bolted on to a platform high above the roof of his house. The house was at the edge of his wood with good sight lines along the rides. Having built his cannon, Hangar bought several huge sacks of marbles – the sort children play with. He then moulded hundreds of clay balls, each approximately the size of the bore of his cannon. He took these to be baked at the local brick kiln, first boring three or four holes in each. The holes were about the diameter of a finger and they were bored right through each ball.

Hangar loaded his cannon with two handfuls of marbles followed by a couple of clay balls. On the first occasion he spotted what he thought was a poacher he fired the massive load. The noise was utterly terrifying – a sort of deafening whizzing, followed by a noise like the rattle of a machine gun as the marbles cut through the leaves and branches of the trees.

Hangar also built his gamekeeper a house on the opposite flank of the wood. The poor old gamekeeper's house had no door or window of any kind on the ground floor. These lower rooms were lit from windows fixed at special angles on the first floor and the front door was situated ten feet above the ground. The idea was that the keeper could draw the ladder up at night safe from any possibility of a revenge attack by poachers. By his keeper's house Hangar built a thirty-foot-tall round tower with another six-pound cannon mounted on it. A walkway connected the keeper's bedroom to the top of the tower.

The idea was that virtually the whole wood was within range of the two guns and, just to make sure that the local poachers were in no doubt as to their intentions, Hangar and his keeper fired continually into the wood day and night. This effectively ruined the shooting anyway, but he was delighted and claimed an absolute success for his invention. 'I may have nothing to shoot in the way of birds,' he is reported to have said, 'but by heaven the poachers know I mean business!'

HOT-AIR HUNTING

ENGLAND, 1839

John Connell, a poacher active in the 1830s and 40s, was a remarkably inventive man even by the standards of a remarkably inventive age. It was almost as if the inventive spirit of the age of railways and engineering had inspired his dim forgotten corner of rural England. He knew that the best time to poach pheasants is when they are roosting, but the wood nearest Connell's home with the richest pickings was also nearest the keeper's cottage. Even a relatively silent .410 shotgun would be too loud. A long stick to whack the birds out of the trees would almost certainly make them squawk and alert the keeper or one of his dogs.

What was to be done? Well, with a friend and fellow poacher Connell began to experiment. If I can't shoot them, he thought, I'll suffocate them. He tried many different chemicals and used chickens to test them. The chicken was placed in a cage in a tree and the chemicals were wafted under the bird's beak on the end of a stick or burned in a bag at the bottom of the tree. The tests were inconclusive: sometimes the bird keeled over, other times it seemed more awake than ever. Only brimstone seemed to work consistently well, but brimstone had to be lit and how was burning brimstone to be carried to the woods and used there effectively?

Then Connell had a brainwave. He got the local blacksmith to make him a three-foot-high stove that could be collapsed like a telescope. It was cylindrical and tapered at the top. The six tubes from which it was constructed fitted into each other perfectly. It had a few holes at the top to let the fumes out and vents to allow a draught of air in at the bottom. Fitted with a brimstone candle at the base, the whole contraption could be collapsed and carried inconspicuously under a big top hat or in the folds of an overcoat. The blacksmith also made Connell two lightweight metal poles that fitted into recesses either side of the body of the stove. These would be used to carry the stove from tree to tree after the metal had become too hot to touch.

The first dark night they tried the poaching stove it worked a treat. After a few minutes under each tree the three poachers who carried it about would hear the soft thud, thud, as pheasants simply dropped like conkers. No one knows how long Connell used his stove, but he was one of the most successful poachers of the age and made enough money from his misdeeds to fund a comfortable retirement.

MUSCULAR ENGINE

ENGLAND, 1839

In the 1830s, when the railway was still young and thrilling, neither the public nor the specialists were convinced that the right system had been hit upon. The locomotives ran on steam; the fact that power that did not involve the horse had been discovered seems to have inspired all kinds of ideas for newfangled and often outlandish modes of transport. One reputable railwayman suggested a 'patent aerial steam carriage which is to convey passengers, goods, and dispatches through the air, performing the journey between London and India in four days, and travelling at the rate of 75 to 100 miles per hour'. All kinds of substitutes for locomotives on the ground were also eagerly sought.

The Globe newspaper reported that a 'professional gentleman at Hammersmith had invented an entirely new system of railway carriage, which may be propelled without the aid of steam at an extraordinary speed, exceeding 60 miles an hour'. No details were given.

Another writer said that the Edinburgh and Glasgow Railway had 'the discernment to employ a Mr Davidson, a gentleman inventor of much practical knowledge and talent, to construct for them an electro-magnetic carriage. The carriage, 16 feet long by 7 feet wide, was duly placed upon the rails, and propelled by eight powerful electro-magnets about a mile and a half along the railway, travelling at the rate of upwards of four miles an hour, a rate that might be increased by giving greater power to the batteries, and enlarging the diameter of the wheels.' This singular contrivance was described as a 'far more valuable motive power than that clumsy, dangerous, and costly machine the steam-engine'.

Perhaps the most extraordinary alternative to steam power was tested in 1839. The Patent Office records for that year mention two gentlemen known as Taylor and Couder who registered an ingenious system by which a carriage was to be drawn along the line 'by the muscular power of the two guards who constantly accompany it'. The carriage was described as very light and elegant in appearance, and capable of carrying seven or eight passengers at the rate of 18 miles an hour. 'We have no doubt,' says a railway newspaper, 'that these machines will come into general use, as they will effect considerable saving to the company in time, trouble and expense of running an engine.'

When there was an attempt to trial the new carriage no two guards of sufficient muscular ability to move the thing an inch could be found and the idea died even as it was born.

THE MOUSETRAP

AMERICA, 1844

For most people the Wild West means cowboys rushing around on horses and firing handguns at each other or being ambushed by Indians in full war cry. There's no doubt that central to the whole history of the Wild West and the pioneers who opened up the frontier for settlement is the revolver, a deadly – if often wildly inaccurate – weapon that could be kept in a holster 24 hours a day, seven days a week regardless of whether the wearer was in bed, in church or dancing a jig.

Revolvers and other handguns were carried by youngsters, priests, elderly schoolteachers and pretty much anyone who could lift and point, and a huge amount of science and skilful engineering went into making revolvers more accurate, more reliable and more powerful.

Among a host of relatively sensible innovations – adding a spiral groove inside the barrel, for example, greatly increased accuracy – there were some remarkably zany ideas.

One amateur scientist in the early 1880s decided that if guns could be set up on trip wires to kill house burglars and others (and they were regularly employed for this purpose) then why not use them in a range of other automatic traps and ambush devices? Hunters would rig elaborate trip wires to rifles and shotguns in an attempt to bring down an easy deer or two, though they often ended up killing themselves or passers-by.

One inventor went so far as to design, build and manufacture a mousetrap that depended for the *coup de grâce* on a magnum handgun. Anyone who has seen the damage such a gun can inflict on a large animal let alone something as tiny as a mouse will realise that this was more a novelty act than a pest control device. But it was carefully built and fully patented. It involved a long rectangular wooden box with the gun fitted in a tight cradle at one end and aimed carefully at the other, open end of the box. When a mouse tried to pull the cheese or chocolate off a spike on a small platform, its efforts released the trigger which had to be set to go off at the least movement – a classic hair-trigger.

By all accounts the mousetrap worked, but only in the mad lawless days of the Wild West could something as crazy as a magnum mouse-killer be left lying around the house for anyone to pick up. A mouse hit by the bullet from the magnum would be almost vaporised but the real reason the magnum mouse-trap fell out of use was that the bullet missed as often as it hit its target!

WALKING TALL

ENGLAND, 1845

Over the past few centuries human beings seem to have become obsessed with physical perfection. The idea of what exactly constitutes physical perfection varies from country to country and from epoch to epoch but there is no doubt that every society worships physical beauty in some form or other. For certain African tribes it is so essential for women to have long necks that special rings are fitted to a child's neck at birth and added to regularly to force extra length into the neck during the growing period. Other tribes cut holes in women's lips and insert increasingly large stones until the lips may hang down for a foot or more. In medieval England red hair was seen as such a splendid thing that any girl lucky enough to be born with it could have the pick of the eligible bachelors.

Today we are obsessed with youth and slimness. Models and other icons of beauty, at least in the West, have to look virtually malnourished to be considered truly beautiful – as Wallis Simpson, Duchess of Windsor, put it, 'You can never be too rich or too thin.'

In Victorian and earlier times in Europe there was also an obsession with straight backs, legs and arms. At birth, babies were swaddled – wrapped tightly in bandages with their legs out straight and their arms kept firmly in line with the sides of the body. They were left swaddled for at least the first three or four months, which must have been torture for a baby desperate to practise moving about. The idea was that a properly swaddled baby would go on to be tall and straight. In Victorian times backboards were invented to achieve the same end.

These must rate among the most bizarre of all beauty aids. Children, particularly girls at expensive schools, were encouraged to walk up and down for several hours each day with their heads held high and two heavy mahogany boards attached to the back and front. The two boards were fixed to each other using leather straps and a vague harness arrangement also went over the girl's shoulders to hold the boards in place. They were made to come right up under the chin in order to ensure that posture – with which the English seem to have had an obsession – was properly maintained. Posture mean an erect carriage and the head held high.

But the crazy inventor of the backboard doesn't ever seem to have considered the possibility that by the time the backboard was recommended for use – when a young woman reached the age of thirteen – she was as straight or as crooked as she was ever likely to be. By that stage no amount of deportment training, using backboards or piles of books on top of the head, was ever going to make the least bit of difference.

BIZARRE BRAKES

AMERICA, 1845

After initial scepticism, the invention of the steam engine, and the speed and ease of transport it offered, captured the imagination of the public as nothing had before. For scientists and engineers the railway was *the* thing to be involved with; many were to devote their lives to creating refinements and variations on what was considered the wonder of the age – a technological marvel that ultimately created an almost religious belief in the power of science and technology.

Despite all this effort, early trains were built with features that seem incredibly batty by today's standards. The guards on branch trains, for example, had to operate the brakes of the trains from the roof of the carriages! They also had to sit for the whole of each journey in a little box built on to the outside of each carriage at the back and high enough above the roof of the carriage to ensure that the guard could see where the train was going; in the little box was a seat but both box and seat were open to the elements and faced the way the train was travelling. This extraordinary invention meant that the guard was buffeted by every breeze, soaked by every downpour and nearly choked by the continual stream of smoke coming from the engine up ahead. Half blinded, often frozen or drenched the guard still had to apply the brake whenever necessary by turning on the hand screw.

One guard remembered a journey across Norfolk in 1844: 'The dust, the smoke, steam, and smother, which filled my eyes, ears, and nose during that short ride, were sufficient to put a stop to any wish for further experience in that direction.' After a while the roof seat was equipped with a small box-shaped shelter for the men, with a small rough curtain added to provide some shelter from the weather. The open van, a vehicle with one end covered, was the next advance.

In America as late as 1845 the system of working the brake on the roof of each carriage was still in force. The 'Brakeman' ran along the top of the vehicles, applying the brakes when necessary – a very dangerous business in rough weather, and especially so on lines with tunnels; in 1845, as a protective warning, overhanging gantries with pendant whipcord pellets were provided not far from the tunnels, so that the men on the roofs, experiencing the warning stroke of these pellets, could lie down while passing through.

CLASS DIVISION

ENGLAND, 1850

The early railway prided itself on the fact that its first-class service attracted the well-to-do, the successful and the more adventurous of the aristocracy. Refinements to the service were always aimed at first-class passengers and at the groups who could be assumed to enjoy travelling in some style. After all, this is where the greatest profits were to be made. And it is easy to imagine the railway owners lamenting the fact that it was necessary, merely to keep the wheels of commerce in motion, also to provide third-class coaches, or wagons as they might better be described.

Third class made little in the way of profit for the railway company owners but factory owners who might have a share or two in railway stock wanted cheap transport for their staff. Imagine then the horror of the railway directors when they heard rumours that the well-to-do were, in increasing numbers, buying third-class tickets.

One director simply could not believe that this was happening so he spent several weeks travelling each day on one of the early up-trains to London. He discovered that a significant proportion of those who, judging by their dress, should have known better, were happily asleep or reading their newspapers in carriages made to carry only those lesser mortals who dug the roads or cleaned the sewers. Those behaving badly included landowners, gentleman farmers and even, God help us, the odd baronet.

In short, first class had invaded third, an event that hinted at the worst kind of revolutionary fervour and, worse, meant a loss of funds for the railway directors and shareholders. This situation could not be allowed to continue, but the solution was hard to find. Booking office clerks could hardly be asked to refuse to issue third-class tickets to anyone who seemed able to afford first class.

The director reported the situation to his fellow directors: 'I discovered that there were certain persons in superior positions who were base enough to travel third class and in order to bring these offenders to a proper sense of their position and to swell the revenues of the company, I recommend that we introduce what might best be termed special inconveniences.'

An extraordinary meeting was held in secret between engineers, company officials and directors. The engineers – hugely inventive to a man – came up with the solution and it was first adopted by the Manchester and Leeds Railway, the directorate of which brought in what was known as the 'soot bag system'.

Thus on a bright spring morning in 1850 began one of the most extraordinary experiments in transport history. A team of four chimney sweeps had been specially hired. Each was assigned a third-class carriage that was believed to harbour individuals who did not belong to the working classes.

The decision about which carriages needed 'special inconveniences' was based on observations made by a senior porter who watched to see if anyone well turned out entered a third-class carriage. He then kept a note and reported his findings to the sweeps.

As soon as the train set off on its journey the sweeps moved into action. They entered the third-class carriages and immediately unfolded several of their recently used sacks and began to shake them out, thus covering everyone in the carriage with a layer of grime and dirt. The working men of course took little or no notice since their work obliged them to wear rough and dirty clothes anyway. The well dressed, on the other hand, were outraged. But what could they do – if they complained to the management they would be told to travel first class since third-class passengers were used to dust and dirt and the railway company could do nothing about it.

The ruse worked; the number of passengers travelling first class rose and, as time passed, fewer of the middle classes risked a confrontation with the soot bag men.

According to an early edition of the *Railway Times* other rail companies found it more convenient to make sure that sheep or even pigs travelled regularly in their third-class carriages!

DOWN A HOLE

AMERICA, 1854

The problems of protecting one's property have exercised inventors throughout history. The Egyptians, keen to prevent the lavish graves of the pharaohs being robbed, lowered huge blocks of stone into position using elaborate engineering systems based on giant sand timers. In the medieval period, iron chests were built with locks of enormous complexity. The locks for house doors have gradually become more sophisticated and burglar alarms are now commonplace in offices, shops and houses.

But the history of anti-burglary devices is also rich in improbable devices. Simon Yorke, the last private owner of Erddig, a huge eighteenth-century house in Wales that is now owned by the National Trust, created his own burglar alarm using tin cans tied together with string. These were then left all over the house, the idea being that any burglar creeping around in the dark would trip on the string and pull dozens of noisy tin cans after him, waking the whole house.

Other domestic anti-theft inventions include a clever system that involves building a staircase with one uneven step. What on earth is the point of that, one might ask. Its effectiveness is based on the well-known fact that when we walk up a staircase we judge carefully the first step and then go on to autopilot since our brains have learned to expect all the steps on a staircase to be the same as the first; if one rise is slightly higher or lower than all the others it will almost inevitably trip up a burglar who doesn't know about it.

But the ultimate burglar alarm was invented in 1854 by an American. The idea involved digging a six-foot-deep hole just inside your front door. The sides of the hole were lined with timber and the whole thing concealed under normal floorboards. At night you primed the trap door (which looked like an innocent patch of floorboards) and a complex mechanism ensured that anyone treading on it released the mechanism and plunged down into the hole. The trap door then sprang back up and locks, thus preventing the burglar's escape.

The difficulties with this invention are obvious – anyone who happened to be in the house quite legitimately who forgot the mantrap could end up spending the night in a dark, cold hole and, of course, children would be particularly at risk. One suggestion was to make the trap-door release mechanism work only when the weight of an adult pressed on it. That would keep all but obese children in the house from the danger of falling in, but burglars might then simply use children to avoid being trapped themselves.

The Victorian idea that punishment for theft should be severe is reflected in a number of suggestions that the underground pit should be lined with spikes or half-filled with water to make the burglar's time till morning that bit more uncomfortable, though that would be nothing compared to the earlier use of trip wires, crossbows and shotguns all rigged to go off if anyone tried to force a window or a door.

Life for burglars was generally far safer when science at last came up with effective and reliable electronic burglar alarms!

SCIENTIFIC HORSE MANURE

SCOTLAND, 1860

Mr Lubbock, an amateur engineer from Dudley in the West Midlands, was travelling on the North of Scotland Railway. The train had stopped at a station and despite a great deal of huffing and puffing the engine refused to get started again.

Mr Lubbock was irritated as he had an appointment two stops further on and hated being late. He went up to the engine and spoke to the driver and the fireman. They explained that they couldn't move because certain tubes in the engine were leaking, but help – they insisted – would be with them soon. Mr Lubbock assumed this meant the arrival by some other means of a spare part or two so he was astonished to see a farmer approaching across the fields with a bucket which turned out to be filled with horse manure.

There was a practice – described as almost universal in the early days of steam – of putting oatmeal or bran or, if these could not be had, horse dung into the boiler in order to stop the leaking of the tubes. The idea was the brainchild of a remarkable, if unrecorded, inventor who clearly had a practical head for practical difficulties.

Mr Lubbock watched as the horse dung disappeared into the boiler and then returned to his compartment. Within minutes the train pulled out of the station as if nothing was amiss. He was later told that using dung as a repair medium was standard practice on engines of all kinds.

BACK FROM THE DEAD

ENGLAND, 1862

The science of death has a long, inventive history. Those incredibly clever Egyptians managed to embalm their dead in ways that preserved the physical remains for thousands of years even though they had only a limited number of chemicals compared to the range of preservatives available today.

More recent highly effective ways to preserve dead bodies include an eighteenth-century practice popular among certain religious sects that involved burial in lead coffins. The idea was that microbes and insects could not survive the high levels of lead contamination and the body would be contaminated by the moisture in the soil acting with the soft lead and therefore survive undamaged. This certainly seems to have done the trick and on one occasion in the City of London work was being done in an old church graveyard when workmen found a coffin near the surface. They were horrified to discover inside the coffin an apparently fresh body dressed in a dinner jacket complete with white shirt and tie. The body and clothes were so well preserved that the police were called since the workmen suspected that the deceased must have died recently. It was only when an alert policeman noticed how old-fashioned the clothes looked that the mystery was solved – the body was actually nearly 200 years old but it had been almost perfectly preserved by its lead coffin.

The Victorians were much obsessed by death. After the death of Prince Albert in 1861 a cult of gloomy self-indulgent morbidity seems to have descended on the country, perhaps inspired by Queen Victoria, who remained in mourning for more than forty years and who seems to have encouraged the whole country – unofficially at least – to wear dark clothes and walk around looking miserable. Running parallel with this perhaps rather theatrical love affair with the sombre was a genuine fear of death, and more particularly of being buried alive.

Other deliberate inventions to do with death and burial include a most bizarre idea thought up by an inventive stonemason in the 1860s. Visitors to Highgate and Kensal Green cemeteries in London even today may see the remnants of a strange invention designed to guard against the risk of being buried on Monday and waking up in one's coffin on Wednesday. A number of tombs were built with a hollow stone column running down into the buried coffin. At the top of the hollow column two or three feet up in the air above the tombstone would be a small bell tower complete with bell. If the deceased happened to wake up after burial he or she would be able to pull vigorously on a chain that ran up the hollow column to the bell, which would ring out, bringing rescuers hotfooting it across the fields.

A number of different coffin alarm systems were created around this time and indeed from then until well into the twentieth century when, in a few cases, nervous relatives had electric alarms fitted to their relatives' coffins.

Despite all the terror and fears, however, there is no record of anyone buried with an alarm pressing the button or ringing the bell!

PLOUGH GUNS

AMERICA, 1862

The American Civil War, like all Civil Wars, was bloody, brutal and stupid. The extent of its stupidity can be measured by the fact that there were countless instances where two members of the same family fought on opposite sides.

By the time the war was already well under way, raids on the north by groups of southern soldiers had become a real problem, with food growing and supply disrupted in a number of areas. It was difficult to arm farmers because so much of their work could not be done while they were carrying a rifle, and a pistol was likely to be of little use to a lone man faced with a whole group of enemy soldiers. Two New York gun makers came up with an answer of sorts. They invented a full-size plough that doubled as a heavy gun.

C M French and W H Fander designed their odd-looking plough gun so that, to the casual observer, it would appear absolutely identical to a real plough. The part that turned over the soil (the ploughshare) was pulled in the normal way behind a team of horses, but the section of the plough that ran forward towards the horses and to which their harness and gear was attached concealed a large-bore cannon capable of firing cannonballs weighing up to 3lbs.

The idea was that at the first sign of trouble the farmer would unhitch his horse team, aim the gun (which had of course been prepared with powder and projectile the night before), and light the fuse. Even a couple of dozen mounted soldiers might turn tail and run faced with the vast explosive power of such a gun, but there was one problem and it was soon pointed out by senior politicians to the plough gun's makers.

Once word got out that ploughs had been converted into heavy ordinance, all farmers would be shot by marauding soldiers as soon as they were spotted. At least with the old real ploughs robbery was usually the worst that happened and though food and livestock disappeared the farmers themselves were not generally harmed.

CALMING ENGINE

ENGLAND, 1863

Dr Walter Lewis, who worked as a medical officer at the London Post Office in 1863, issued a report on the possible dangers of rail travel. He explained that after a long period of research that included numerous questionnaires, medical examinations and psychological analysis he was able to reassure the public that 'railway travel has little, if any, injurious effect on healthy, strong, well-built persons, if the amount be not excessive, and if they take moderate care of themselves; but that persons who take to habitual railway travelling after the age of 25 or 30 are more easily affected than those who begin earlier, and that the more advanced in age a traveller is, the more easily is he affected by this sort of locomotion. Weak, tall loosely-knit persons, and those suffering under various affections, more especially of the head, heart, and lungs, are very unsuited for habitual railway travelling and should avoid it unless absolutely necessary.'

Dr Lewis also explained that his researches had enabled him to come up with a cure for those in these high-risk categories who really had no choice but to travel regularly on the railways. His patent traveller's calming engine was, he said, 'the only certain palliative for the potentially severe health effects induced by high speed rail locomotion'. It had been designed and manufactured by the 'first scientists in the land'.

And of what did the calming engine consist? It comprised a wooden box containing smelling salts, a small bottle of brandy, a cap that came down over the eyes and the palliative pressure pad – which looked like a giant pair of brass scissors except that instead of blades it had two flat pads. These were placed on the temples when the pressures of rail travel threatened to become too much too bear. The handles of the 'scissors' were then squeezed, which pushed the pads down hard on the temples. The result – provided you'd already taken a good swig of the brandy – was instant relief, a sense of calm, and a renewed vigour which would enable the traveller to continue the journey unaided and without the need for more formal medical assistance.

GOING ROUND IN CIRCLES

RUSSIA, 1873

Ships, tanks and aeroplanes designed for war can always be improved, but what looks good on the drawing board doesn't always work in the real world – as the Russian Admiral Popov discovered in 1873. That was the year that saw the launch of one of the strangest military machines ever invented.

It was early one autumn morning that the warship *Novgorod* put to sea from the Black Sea port of Nikolayev. Rumour had been rife for months among the inhabitants of the town that this was to be no ordinary launch and as the ship hit the water there were gasps of astonishment. At more than 2,500 tons the *Novgorod* was big for her time; she was also bristling with armaments – and perfectly circular. Popov had been faced with a difficult design problem: he needed a warship that could operate in shallow coastal waters, but he had been given a lengthy list of armaments by the Czar and was told that everything on the list had to be on the ship. The Czar, who was paranoid that his country was in danger of invasion, had put so many things on the list that a conventional warship small enough to do the work of patrolling inlets and bays would have sunk. The *Novgorod* was the solution. It had a shallow draught and, being circular, could achieve angles of fire from its myriad weapons that were only dreamt of on conventional warships.

The *Novgorod* had a central tower fitted with two massive ten-ton guns that could be turned full circle; the ship was kitted out with no fewer than twelve powerful screw propellers and, properly used, they could propel the ship forward or backward or even make it spin rapidly clockwise or anticlockwise. The sailors who manned the *Novgorod* reported that it was unusually stable even in rough weather.

The biggest drawback to the circular warship, and the main reason only two were ever built, was that it was very slow. With a top speed of about ten knots it could never compete in open water with more conventional ships and with no keel and a very shallow draught it was incredibly difficult to steer – one captain reported that steering the *Novgorod* was like ice-skating with butter on the soles of your boots!

PARACHUTE HAT

AMERICA, 1879

In 1879, a Mr B B Oppenheimer, worried at the increasing number of high-rise buildings in American towns and cities, bent his not inconsiderable intellect to creating a rescue system that would be cheap and effective. It would also depend on each individual choosing to adopt the system if he or she was sufficiently worried – those who were not worried would simply not buy Mr Oppenheimer's device. This would have the great benefit of saving costs for the companies who built the tall buildings since they would no longer have to install fire-prevention devices, which were pretty primitive then anyway.

The Oppenheimer Fire Escape was very simple: it looked exactly like a small circular upturned boat and was attached to the wearer's head via a leather hat. The idea was that when and if a fire started you placed the boat-hat on your head, tightened the straps under your chin and walked out through the nearest window before floating gently to the ground.

As anyone who has done any parachuting will know, creating just the right amount of controllable lift is not actually that easy – which is why parachutes are so carefully designed – and it must have quickly become apparent to Mr Oppenheimer that his upturned boat idea was simply not going to work. Although it was built of canvas on a wooden frame it was still too heavy and too small to do what was expected of it – Mr Oppenheimer also failed to take into account that the people he was trying to protect were all different sizes and what might work for a small man would certainly not work for someone weighing fifteen stone. It is extraordinary, too, that in the surviving design for the Oppenheimer Fire Escape system there is nothing to indicate that Mr Oppenheimer himself had any awareness at all of aerodynamics or the physics of parachutes. The boat-hat is simply an upturned dish and we must be grateful that it did not get further than the drawing board.

IN AT THE DEEP END

AMERICA, 1882

With the Victorians' prudish reputation it is hard to believe that they were keen swimmers. In fact it was the Victorian passion for the seaside – 'all that glorious ozone!' as one bishop put it – that created places like Weymouth and Brighton. Brighton had been popular in Georgian times but bathing machines along the beaches here and elsewhere along the coast began to appear in numbers only when Victoria ascended the throne.

The passion for sea bathing had much to do with the idea that exercise created stout-hearted Englishmen and women with stiff upper lips and able to go out into the world and run the Empire. The Victorians also felt that bathing improved the moral fibre of the swimmer – by that they probably meant that regular immersion in icy water reduced the chances of people having impure thoughts or indulging in impure actions (in modern language that means thinking about sex).

The bathing machine was, in fact, a most bizarre contraption – its main function was to protect the bather from prying eyes eager for the sexual thrill of a pair of ankles. Bathing machines were wooden carriages roofed over and fitted with steps on the seaward end and a set of harness chains at the other. The chains were attached to a horse and the bathing master or his assistant worked the horse backwards to lower the bathing machine down the slope of the beach and into the sea. Once the machine was half submerged, the timorous bather (dressed only in a stripy neck-to-ankle bathing suit) tiptoed down the steps and into the sea. When he or she decided enough was enough, the procedure was reversed: the swimmer clambered back up the steps, the bathing master geed up his horses and the machine was pulled back up the beach and out of the water. Inside the sealed wagon the bather would get dressed and appear decorously twenty minutes later. Some bathers climbed aboard privately owned wagons at home and had themselves pulled to the sea and back again to reduce still further the risk of being seen.

Bathing machines were popular in Australia and the United States during the same period but the inventive Americans decided that they could improve the lives of keen bathers with a number of variations on the bathing machine theme.

Perhaps the most extraordinary was the machine invented by two New Yorkers – Lorenzo Dows Smith and John Buster Smith. They fixed a large wooden cross-shaped structure to the sea bed about a hundred yards off-shore. In position the cross was just submerged. To it were secured four or five strong cables that fanned out as they ran back along the surface of the sea to the shore and a series of stout posts to which they were fixed. The cables were tightened until they ran just above the water all the way out to the cross. A short running line was then attached to each cable. At the bottom of each line was a harness that could be worn by a swimmer. You hooked yourself on to a line, walked into the sea and made your way out towards the wooden cross. On the way you could swim if you were able (and wanted to exercise) or you could simply float and paddle a bit. The two Smiths were convinced their machine was a winner because it would mean that both good swimmers and complete novices could swim in the certain knowledge that they couldn't get into difficulties because the running lines were short enough to allow the swimmer to grab the main cable if they were tired or worried or suffered a sudden bout of cramp. As soon as the operator (based in a lookout tower back on the beach) saw someone grab the main cable he used a special cranking device to haul them to shore.

For the really nervous, the Smiths attached a floating basket to the running line and dragged the lazy bather up and down – a sort of primitive Jacuzzi!

DIMPLE MAKER

ENGLAND, 1884

Anything that seems to offer people the chance to be taller, healthier or better looking is likely to be snapped up by the vulnerable and gullible. At one end of the spectrum shampoos with peach nut oil or essence of ginger are pretty harmless and, though there is not a shred of evidence that such preparations are any better than soap and water, we at least don't put ourselves in danger if we try them.

In Victorian times the laws governing advertisements were very different and the sort of companies that now advertise harmless if silly additions to their cosmetics would, a century or more ago, have been able to get away with far more extravagant claims. Take the inventors of the Dimple Maker. This bizarre device pandered to the fashion for dimpled cheeks – a fashion as irrational as any other fashion but none the less real for all that.

The Dimple Maker came as a carefully boxed kit that was guaranteed to produce a dimple in double-quick time – it did it, but in the most gruesome fashion.

The kit consisted of a long, thin miniature scalpel with a razor-sharp blade, a miniature (and equally sharp) thin-handled spoon, and a needle and some silk thread.

The instructions explain how these curious instruments were to be used.

The dimple-desperate would use the knife to make a tiny but quite deep cut in the cheek. Into this cut the sharp spoon was then quickly inserted (presumably you had to be quick in case you fainted as the blood coursed down your cheeks!) and turned in such a way as to gouge out a dimple-shaped lump of fat. Once you'd cut out the fat you simply sewed up the wound with the needle and thread.

Nowhere in the instructions for this device (first marketed in the 1890s) does it mention pain or the risk of infection. It is also very difficult to believe that the effect of the dimple maker would be anything other than a nasty scar; a scar moreover that bore no relation to a dimple.

The makers, however, quoted one Mrs R J Rentlesham as saying that the dimple she was able to create for herself was 'as effective as the genuine print of an angel's kiss'.

EAR FLATTENER

AMERICA, 1885

It has often been said that every human, on examining themselves in the looking glass, finds something they long to change; it might be skin tone, or eye colour, but anyone teased for having big ears at school will know that ears that stick out are the bane of life and anyone who suggests that they might be able to do something about it is likely to be welcomed with open arms.

Thus when Fred Haslam set himself up in business in America in the 1880s he homed in on that percentage of the human race afflicted with over-sized hearing equipment. He produced the grandly titled Fred Haslam and Co Apparatus to Prevent Projecting Ears. In two sizes ($1.50 each) the ear flatteners consisted of a leather strap that ran under the chin and over the head with a further strap attached to the first strap about halfway up the face and running to a buckle behind the head. Just behind the point where the two straps met an oval-shaped stiff rubber ring was attached. When the whole thing was strapped to the head the ovals on either side lay on top of the ears, pressing them flat against the head.

Advertisements for the Haslam Ear Apparatus solemnly announced that the device could be comfortably worn during sleep or while working, gardening or doing household chores.

We don't know how long the Ear Apparatus remained on sale but by all accounts it was popular. Didn't anyone notice that when you took it off your ears were just the same?

LOVE HANDLES

ENGLAND, 1886

The Royal Train built for Queen Victoria was fitted – by her special request – with a specially designed double loo. This was created to allow the Queen to pop to the smallest room with her beloved Albert if either happened to be caught short during a particularly interesting conversation. If they went to the loo together the conversation could be continued – vital for Victoria as Albert was the great love of her life and, as Victoria herself admitted, they shared everything.

The Queen also delighted in love seats – a sofa with two seats designed in a vaguely s-shape so that lovers could sit side by side and back to back as it were. The idea was that the lovers' heads would be sufficiently close together to allow for intimate conversations, but their bodies kept well apart – ensuring that there would be no physical impropriety.

Romance, at least of the spiritual kind, was a big thing in Victorian times and these and other inventions made manufacturers rich – which explains why so much time and energy was devoted to finding additional products that would appeal to the romantically minded. Scientists and inventors knew that where romance was concerned there was money to be made. Among the more fanciful of their inventions was a hat with a glass frame built beneath the front of the brim; here a lock of the loved one's hair could be placed and looked at longingly while the hat-wearer was in transit. Then there was the two-handled love cup, a sort of miniature china version of the love sofa.

Perhaps best of all were the lovers' gloves. When you bought a pair of these what you actually got was a pair of gloves with room for three hands. If you wanted to hold your girlfriend's hand in your right hand you bought the right-handed version; if the left, then you bought the left-handed version. In both cases one glove was fitted with two sets of fingers (and thumbs) and the idea was that your girlfriend slid her hand into one set of fingers while you slid your hand into the other. The double glove had only one palm area, where the lovers' hands would meet as the couple strolled along. The great benefit of the gloves was that you could hold hands and avoid the huge embarrassment of being seen in public with naked hands – for no decent Victorian would venture out of doors without hat and gloves.

HAIR-RAISING

ENGLAND, 1887

Facial hair is now generally unfashionable yet a century and more ago most adult males in England would have sported either a moustache or a beard or both. The Victorian attitude to an item of facial hair was like the Victorian attitude to a hat – you should never leave home without it.

The problems large beards caused in an age much concerned with etiquette must have been considerable – for the very hairy there is a point at which eating and drinking become decidedly messy and unpleasant, at least for spectators. This was a particularly aggravating problem for those who spent years cultivating long, elaborate and sometimes very bushy moustaches.

At least two bizarre inventions tried to cater for the needs of the moustache enthusiast who wanted to keep his facial hair but still enjoy the social round without embarrassment.

The moustache trainer – the product of a now forgotten late-Victorian inventor – was designed to keep the ends of a long moustache well up and out of harm's way. It consisted of a hat made from thick strands of wire that criss-crossed the head but left wide gaps through which the dinner guest could pull his hair in order at least partially to conceal the fact that he was wearing a hat. From each side of the hat somewhere near the ear two thicker strands of wire came round the face and ended in two narrow brushes. The brushes were designed to hold the bristles of a long moustache in position, well away from the face and/or the dining table.

The wearer simply donned the hat and sat down to dinner. Just before the first course arrived (or perhaps up in his bedroom before the meal according to preference), he would lift first one side of the moustache and then the other and coil the ends round the brushes. After dinner the ends of the moustache could be unwound and the wearer could set off for the drawing room, his facial hair intact and unsullied.

Another manufacturer came up with the idea of a moustache protector that looks rather like a semicircular curved piece of smooth metal with an oval hole in the middle. This moustache protector was carried by the moustache wearer wherever he (or conceivably she) went. As soon as a cup of tea or a soup spoon hove into view he would take out his moustache protector and attach it to the edge of cup or spoon and drink through the small hole, the metal flange round the hole in the protector ensuring the pristine survival of the facial hair.

BIRD-DRIVEN BALLOON

AMERICA, 1887

Animals of all kinds have long been used to provide a means of propulsion – from horses and donkeys through elephants to goats, sheep, llamas and dogs. But the strangest animal-propelled invention of all has to be the Victorian Lord Rothschild's zebra carriage. He spent years training two zebras to pull this elaborate contraption through the streets of London although the experiment was only a partial success as the zebras were far more inclined than donkeys or horses to stop dead in the street or kick over the traces.

Other unusual attempts to harness animal power include a remarkable idea to use horse power to propel boats. The idea was based on the fact that donkeys and horses were often used to drive mill machinery or to pull buckets of water from wells. The horses were fitted to a treadmill which turned a screw propeller. The only difficulty with the idea – sadly it was an insurmountable one – was the relatively small amount of power generated by the horses compared to the amount of feed they needed and the space required to house them.

An 1887 patent for a horse-drawn boat makes more sense simply because it is a small boat and the horse swims as it pulls rather than riding on an elaborate energy transfer system which is what the treadmill is. The patent drawing shows the horse pulling what looks like a speedboat and what makes it so wonderfully crazy is the presence on each of the horse's legs of a paddle. These broad blades are attached at the fetlock and the aim clearly was to try to turn each of the horse's legs into a kind of oar. What the inventor failed to remember, however, was that though the amount of propulsion per stroke would increase because of the blades, the horse would quickly become exhausted because of the extra effort involved in moving its legs against the dense water.

An even more fanciful idea was also patented in America in 1887 – it was for a hot-air balloon with a difference. Before fixed-wing aircraft, balloons were certainly more than capable of reliable extended flight but they had one major disadvantage – they had to rely on wind to move them. They could propel themselves up and down but not laterally. The solution according to one American scientist was to attach a number of eagles, vultures or other large birds to a balloon. The eagles would be on the end of some sort of harness arrangement; with half a dozen of these large birds all flapping away together the problem of balloon propulsion should be solved. No one seems to have considered how the eagles could be persuaded to fly continually or for any length of time at all. And the means by which they were to be encouraged to fly in particular directions is still a mystery – perhaps the pilot of the balloon would have a long stick with a tasty morsel dangling from the end. If held out just beyond the reach of the eagles in the correct direction it is conceivable the eagles would fly in endless pursuit of the unattainable titbit. History is silent on this question but we know that the eagle-powered balloon never really got off the ground. The best description of the eagle-propelled balloon should come from the inventor himself. In his patent application he explains:

> To all whom it may concern:
>
> Be it known that I, Charles Richard Edouard Wulff, of Paris, in the Republic of France, have invented a new or improved Means and Apparatus for Propelling and Guiding Balloons (for which I have obtained Letters Patent of France for fifteen years, dated April 21, 1886, No. 175,662) and I do hereby declare that the following

is a full and exact description thereof, reference being made to the accompanying drawings.

All attempts heretofore made to guide or steer balloons have comprised mechanical, electric or other motors for imparting the necessary speed and direction to the propelling parts. Attempts in this direction have generally been unsuccessful by reason of the weight of the motor and its accessories, and because the propelling and guiding parts are only imperfectly suited to the medium in which they are placed and against which they have to take their bearing. By this present invention the mechanical motor and propelling and guiding arrangements are replaced by a living motor or motors taken from the flying classes of birds such as, for example, one or more eagles, vultures, condors, &c. By means of suitable arrangements all the qualities and powers given by nature to these most perfect kinds of birds may be completely utilised.

The balloon may be of any suitable form and dimensions. It should be adapted to fulfil in the best possible way all the conditions indicated by theory and experience for obtaining the maximum speed of translation and perfectly stable equilibrium. The gas chambers are placed in front and behind. Their sizes are determined by the number of persons to be carried, and by the specific gravity of the materials and accessories entering into the constitution of the balloon.

Supports or stands rise sufficiently high to be connected by hinges with four corsets, in which are secured by bands and shoulder straps the birds intended to draw and direct the balloon.

The corsets or harnesses have forms and dimensions appropriate to the bodies of the birds chosen, such as eagles, vultures, condors &c. The straps secure the birds firmly, but leave their wings in perfect liberty. The corsets are capable of pivoting forward or backward. The aeronaut cranks a handle to orientate the balloon such that whatever way the birds fly their efforts will be directed in the way the balloon driver requires.

The result of these arrangements is that the flight of the harnessed birds must produce the motion and direction of the balloon desired by the conductor, whether for going forward or backward, in a right line, to the right or to the left, or for ascending or descending.

It may be observed that the birds have only to fly, the direction of their flight being changed by the conductor quite independently of their own will.

When the apparatus stops, the birds rest on the stages. A net or nets is then lowered to prevent the birds from flying.

The balloon should always remain as much as possible in stable equilibrium, both when on the ground and when floating in the air. No weight should be left to be supported by the harnessed birds, so that the whole of their flying power may be utilised for advancing or guiding the balloon. A store of ballast enables the conductor to obtain this result constantly.

A BICYCLE WITH A DIFFERENCE

ENGLAND, 1889

Hugely creative, and convinced that technology could solve every problem, the Victorians came up with dozens of gadgets that, in a more evolved form, are still with us today – the telephone and the vacuum cleaner to name but two. They also invented the bicycle, of course, but not content with the basic bicycle they tried to come up with modifications and improvements that were sometimes taken to absurd lengths.

Anton Oleszkiewicz, who was said to be far more English than the English despite his Russian surname, was convinced that the velocipede, as the bicycle was then known, could be vastly improved if the cyclist's every movement – and not just the movement of his or her legs – could be used to power the machine.

And so it was that in the autumn of 1889 he announced to the world his Winged Messenger or New Improved Driving Mechanism for cycles. This consisted of a complex series of levers and springs that ran from the back wheel of the bicycle via the saddle and all parts of the frame to an elaborate, all-encompassing leather harness that was fixed around the cyclist's upper body. One cyclist who tried it out took half an hour to strap himself in; once that was done he then had to attach himself to the first of the levers – a long steel rod that ran from a position on the harness at the front of the chest down to a horizontal rod that was hinged, spring-loaded and fitted to a chain wheel just below the saddle.

To ride this extraordinary new bicycle the rider, having strapped himself in, was advised to pedal in the normal way until he had attained 'a stately speed'. According to the instructions he was then to begin throwing his upper body backwards and forwards 'rhythmically and with as much violence as possible'. Each movement was, in theory, transmitted down the chest rod into the hinged lever and on to the small chain wheel above the back wheel. An extra chain ran from this wheel down to the main chain wheel in the centre of the cycle's rear wheel.

Oleszkiewicz organised a trial day in London's Hyde Park and passers-by were no doubt astonished to see a group of cyclists apparently suffering from seizures as they whizzed along by the Serpentine. The experiment was not a success, however, as two cyclists fell off the back of their machines, several suffered extreme motion sickness and none could discern any increase in their speed. Nothing daunted, Oleszkiewicz went on to try his hand at man-powered flying machines.

MOVING BABY

AMERICA, 1890

From giant prams and baby carriages we've come a long way to today's apparently simple folding pushchairs but what looks simple is actually deceptively complex – engineers and scientists spent years trying to improve the way we carry our children around.

As in so many fields of scientific endeavour, baby carriages have thrown up numerous wacky ideas and designs over the years and among the best – or worst – there are some real oddities, including the 1920s pram that looked like a miniature tank complete with armour plating and caterpillar tracks. Its inventor thought it would encourage more women to push their babies across rough terrain and through woodland. Falling branches, avalanches and rock falls couldn't kill baby because of the sturdy metal construction of the pram, and the caterpillar tracks ensured smooth running through bogs, over sand and gravel, and even through boulder-strewn territory.

Even odder than the tank baby carrier was George Clark's baby carriage that imitated a large shoe! Clark was an inventor and entrepreneur who decided that the novelty of a pram designed to look exactly like a giant high-heeled shoe would take the market by storm. His shoe buggy came complete with laces (they pulled the waterproof covering together over the baby) and a built-in umbrella.

To the complete bafflement of the ever-optimistic Mr Clark, his giant shoe baby carriage never really took off and even the few prototypes he made have long vanished.

WATCH THE BIRDIE

ENGLAND, 1890

After the invention of photography in the early part of the nineteenth century – an invention variously attributed to Louis Daguerre and William Henry Fox Talbot – the science of picture-taking moved on rapidly. From a process that could produce only a single unique image (rather than a negative), photography quickly advanced until it had all but ousted traditional portraiture as a means of capturing a likeness. True, the very wealthy and the aristocracy still patronised painters but for the middle classes and increasingly for the working classes the photographic studio was the place to go to celebrate weddings, births and other important anniversaries. By the late nineteenth century cameras were relatively sophisticated, though still primitive by today's standards.

The photographer would carry a large plate camera – the plates were made of glass and could be used to produce any number of copies of the original image – and artificial light could only be generated by burning a strip of highly volatile magnesium. That said, every week brought some new technical improvement. Many of these improvements lasted for many years; others were imaginative and bizarre, to say the least.

Perhaps the most amusing of all photography-related inventions is the artificial bird that was fitted to many studio cameras in the nineteenth century. Photographers wanted a spontaneous look once it was no longer necessary for the subject of a picture to stay absolutely still for half an hour while the picture was taken. But how could they achieve this? Telling people to smile inevitably produces a range of expressions from rictus grin to psychotic stare.

Then some long-forgotten photographic genius came up with the birdie idea. It sounds odd and must have sounded even odder at the time, but it was a good idea. The first birdies were simply attached to the top of the camera. In the instant before the photographer pressed the shutter he would shout, 'Watch the birdie!' and any children in the group would in theory look up at the artificial bird with the light of wonder in their eyes. It worked for some adults too, but then one photographer thought he would go a step further and introduce a real surprise, a surprise that would produce a spontaneous look of delight on the faces of his subjects. To do this he fixed up a kind of jack-in-the-box on a pole next to but slightly above his camera. Inside the box was hidden a small artificial bird with brightly coloured plumage. When the time came to take the picture the photographer would shout 'Ready!' and then press a trigger that released the spring-loaded bird. The instant the bird leapt into the air the photographer took his picture and found the frame filled (in theory) with looks of delighted wonderment.

DUNG CATCHER

ENGLAND, 1892

Before the coming of the motorcar the streets of most cities were awash with horse dung. London, the busiest, most populous city in the world in Victorian times, was probably the worst afflicted. For some the tons of horse dung strewn all over the streets was a nightmare – a nightmare of dangerous road crossings (horse dung is notoriously slippery!) and appalling summer smells – but for others the rich aroma of the countryside drifting through the town was a thing of romance.

Whatever your view, horse dung and its production was an indelicate matter from which the gentler sex really ought to be protected – a belief that led to some serious work among London's scientists and inventors. And by the mid 1890s they had come up with the answer: Calantarient's Improved Dung Trap for Carriage Horses Employed by Ladies of Fashion.

This grand-sounding name conceals a fairly straightforward but ingenious device for protecting delicate females once they'd clambered into their carriages. The problem was that the only view from many carriages – Hackneys, Broughams and so on – was out the front and over the horse's back. So if the horse had to do what nature intended it to do, the poor delicate creature in the carriage was shocked beyond measure. Calantarient's Improved Dung Trap provided the answer by adding a specially designed bag to the carriage harness. When the horse did something unmentionable the dung disappeared into the specially designed bag. A later version had a special lanyard that could be pulled by the driver to release (discreetly) the contents of the bag on to the street below the carriage if, at the end of a long shift, it had begun to fill up.

It was a device that spared the blushes of the ladies and therefore met with huge success in a society obsessed with the idea that women must be protected at all costs from anything in the least indelicate.

IF YOU WANT TO GET AHEAD . . .

ENGLAND, 1894

Advertising really took off in the second half of the nineteenth century. Anyone who has studied early photographs of London's Fleet Street or the Haymarket will have noticed the profusion of billboards, hoardings and posters on every available surface. Sandwich-board men – those poor souls condemned to tramp the streets in all weathers with advertisements strapped to them back and front – were also a feature of the streets of most of our cities.

With so much advertising competing for the attention of passers-by, the hunt was on for something that would really give individual advertisers the edge. The theatrical manager Sidney Squires and his friend the engineer Edward Moorhen thought they had come up with the answer – a device that would make billboards and posters seem decidedly old hat.

The two men invented the Improved Pneumatic Advertising Hat. This consisted of an extra tall top hat fitted with a hinged crown or lid. Beneath the lid in the space between the top of the hat and the wearer's head was a battery and a delicate mechanism attached to a pneumatic tube which ran down the inside back of the wearer's shirt, around his body and out into a large rubber bulb held in the hand. When the bulb was squeezed vigorously and rapidly the crown of the top hat slowly rose until it was at right angles to the wearer's head.

This is where the battery came into its own – as soon as the crown had reached the correct position it lit up with whatever advertising slogan or device had been attached. In principle, of course, this was an excellent invention – who could fail to notice when an apparently ordinary hat on a respectable passer-by came to life, sometimes shooting bolt upright and lighting up, at other times – if the skill of the bulb squeezer was up to it – bobbing up and down as he marched along. Initially the hat sold in reasonable numbers but those employed to wear it complained that children threw stones at them and the weight of the contraption gave them headaches. Within a few years the Improved Pneumatic Hat had vanished into the long oblivion of history.

HORSE-POWERED FLIGHT

ENGLAND, 1895

The range and sheer quantity of man-powered flying machines almost beggars belief. From Leonardo da Vinci's extraordinarily prescient helicopter drawings to Victorian bicycle-driven machines, flying seems to have captured the imagination of artists, writers, scientists and inventors in a way that nothing else quite matches.

Among the most way-out of the early flying machines was an arrangement of two sets of wings, one mounted above the other – the surviving drawings reveal a complex creation of leather, silk, string and wire. On the day of the launch a local press reporter explained that the double-winged bird was almost too heavy for the inventor to stand up in once he'd been strapped into it. He apparently staggered forward a few yards before toppling over and breaking two of his four wings.

The bicycle seems to have inspired renewed interest in man-powered flight, probably because the 15–20 miles an hour achievable on early machines seemed the final piece in a jigsaw that had bedevilled scientists for centuries. Pedal power allied to lift had to be the answer. Pedal-powered machines with as many as eight sets of wings were launched from flat and sloping fields, off cliffs and piers, into headwinds. Nothing worked.

A mid-Victorian book on flying and ballooning mentions one Jean Francois Mercadante who should be awarded the prize for the most eccentric flying machine ever created. A mathematician of some ability, Mercadante calculated that a flying machine that consisted of a tricycle suspended beneath a pair of thirty-foot-long wings (making a total wingspan of 60 feet) and pulled downhill to a fifteen-foot-high cliff by two thoroughbred horses would certainly be the thing to put a man into flight.

He waited months until weather conditions were just right – a steady headwind blowing up the chosen hill, dry ground, not too hot or cold and good visibility – before making his attempt. The grass on the hillside was carefully cut and bumps and irregularities in the runway were either flattened or removed. Two fine stallions were attached by 150-ft ropes to the front of the tricycle. Two men were employed – one to hold each wingtip and to run alongside until speed, wind and lift were sufficient to keep the whole aeroplane on an even keel.

Amazingly the first part of the experiment went rather well. The horses took up the slack and began to pull the plane along slowly. They were then encouraged to trot and the plane's speed increased. The wing holders later reported that they did feel some lift at this stage. The horses broke into a canter, the wingtip holders let go, and the plane increased its speed and reached the cliff edge as the horses veered to one side.

Perfect. Except that in these early days no one was able to calculate speed, lift and weight and the necessary ratio between them. Mercadante's wings, though large, simply didn't produce enough lift to carry tricycle and tricyclist. Mercadante went over the edge of the little cliff and dived straight to the ground. Luckily, he wasn't injured but his giant albatross was wrecked and it was to be several decades until engine power and superior aerodynamics were to solve one of science's most challenging problems.

MUSICAL LOO

GERMANY, 1895

At the end of the nineteenth century musical boxes were all the rage. When they think of musical boxes most people think of small boxes that open to reveal a twirling ballerina and play one or at most two tunes. Victorian musical boxes were very different. Typically they were two or even three feet long, perhaps a foot deep and nine inches high. Inside, a brass cylinder ran from one end to the other. Along the brass cylinder were hundreds of tiny brass pins. When the machine was wound up (using a large folding handle at one end of the box) the cylinder turned and a vast bank of perfectly tuned metal notes (rather like miniature tuning forks) were flicked by the brass pins. According to which tune was selected, using a lever on the outside of the box, different metal notes were struck by the revolving pins. Some of these boxes could play upwards of twenty tunes, each depending on the exact position on the cylinder of the right pins spaced perfectly in relation to each other. The resulting music is a delightful merry-go-round sound of tinkling, but as the boxes grew more complex, ideas for increasing their use were touted widely among inventors.

Perhaps the most incongruous of these notions was a set of lavatory cisterns fitted with mechanical music boxes. At the end of the nineteenth century grand ornate public lavatories were enormously popular in England. They represented a departure from the Georgian and early Victorian period when people still commonly urinated openly in the street simply because there were no facilities. As confidence and national wealth grew, the great clean-up of major British cities began. In London, Joseph Bazalgette built the embankment along the Thames largely to house a vast new sewer that would take London's effluent miles downriver before releasing it into the tidal reaches of the Thames. Before the embankment sewer, sewage went straight into the Thames in central London and as a result cholera was rife; the stench was so bad in summer that chlorinated sheets had to be hung against the windows of the Palace of Westminster or MPs simply gave up and went home.

The public lavatory was a major part of the great clean-up spirit and all over London they appeared, usually below ground. The urinals themselves were massive and ran in six-foot-high ranks along the walls in the men's loos. Made from vitreous china, they were often beautifully decorated and much of the plumbing was of polished copper. The walls tended to be covered with decorative tiles from floor to ceiling. In the ladies' the situation was comparable, with large cubicles with beautifully made mahogany doors and fittings and generous-sized sinks.

The public loo reached the height of sophistication, but the most evangelical lavatory engineers were still not satisfied – which is presumably why a design was eventually submitted for the musical cistern. The plan was that when the cistern for the urinals flushed automatically it would trigger a large mechanical musical box and play a series of tunes one after the other. Then, after a five-minute break and while the cistern slowly refilled, it would be set going again by the next automatic flush. The idea was tried out in a number of London's grandest lavatories but soon abandoned. The music attracted too many people who didn't want to go to the loo at all but were simply enjoying a bit of free public entertainment.

However, the idea of a musical public lavatory never quite died away – in the 1980s a public loo in London's Covent Garden became famous after its attendant began to play records of light Italian opera! So popular was the musical Covent Garden loo that the attendant won a number of awards.

THE WONDERDRUG

ENGLAND, 1897

The discovery and development of the aspirin is one of the strangest stories in science but despite its strangeness it produced a drug of enormous benefit to mankind – so much so, that it has been called the wonderdrug. Apart from curing headaches it now has a vital role in preventing strokes and heart attacks and may even help treat some cancers. But the humble aspirin had quite remarkable origins.

There is some evidence that the Romans either chewed willow bark or made some sort of infusion from it when they needed to treat fevers – they had no idea why it helped but help it did. Hippocrates, writing in around 400BC, mentions a tea made from the leaves of the willow as an excellent treatment for fever.

Scientists in the nineteenth century managed to isolate the chemical known as salicin, which is synthesised to create acetylsalicylic acid, the basic ingredient of all aspirin. That such a vital health tool should come from such a common plant led to fears that continue to this day that rare plants, particularly in the Amazon basin, may become extinct before we get a chance to discover if they, like the willow, can help in the battle against disease.

The first pure form of aspirin was highly effective but in most instances it was too powerful and caused stomach pains. A French chemist called Charles Gerhardt tried mixing salicylic acid with other compounds to reduce the side effects but his process was complex and expensive.

Then, more than half a century later, the German scientist Felix Hoffman found a more effective way to reduce the harmful effects of aspirin while retaining its benefits. The rest, as they say, is history.

In sheer weight we now take over 70 million pounds of aspirin, and in America alone aspirin sales have reached $15 billion a year.

Apart from Viagra – for obvious reasons – it is the world's most popular drug!

HOT HEAD

ENGLAND, 1898

For centuries scientists have grappled with the mysteries of baldness. Various cures have been tried: in the ancient world chicken's blood was dripped over the hairless heads of wealthy merchants in a fruitless attempt to reverse the cruel trick of nature; two thousand years later in the 1960s equally unlikely cures were being tried for the folically challenged including – according to a report in the *Mirror* newspaper – burned toast.

A correspondent to that newspaper wrote to explain that his family had long held the secret of reversing male baldness. All you had to do was burn two slices of toast each day, grind up the carbon remains, mix with a little water and then smear on the afflicted head. New hair should start growing within a week or two. The most astonishing thing about the toast cure is that several readers wrote in to say that they had tried it and it worked. Several scientists wrote to the paper saying that baldness was entirely genetic and that burned toast could not possibly have had any beneficial effects.

However, in the history of cures for baldness one particular remedy stands out – the Kehoe and Jackman Electrical Scalp Stimulator. Surviving illustrations show that the stimulator looked like a medieval instrument of torture. It consisted of a leather cap into which ran half a dozen wires. The bald person strapped the hat tightly to the head and then attached the loose ends of the wires to a wooden box about nine inches square and fitted with a handle. Inside the box a simple gear system ensured that when the handle was turned two 'friction pads' rubbed furiously against each other. The static electricity thus generated was – supposedly – carried up the wires into the cap and on to the hairless head.

The leaflet supplied with every Stimulator explained how the device should be used:

> Painless and highly effective. Use daily either when rising or when retiring to bed. Can be self operated or – for more rapid hair growth – contact our head office for the address of a qualified technician.

When the Stimulator first appeared on the scene, electricity was for most people a new and exciting phenomenon. It seemed to possess unlimited possibilities – medical, industrial and domestic. Though why anyone should think that applying static electricity to the head should automatically reverse hair loss is a mystery. Perhaps it is simply that when we see something in print we tend to believe it!

DRAGGED TO BLACKWALL

ENGLAND, 1899

When the very first passengers boarded the London and Blackwall Railway they were about to travel on what even today must count as one of the oddest railways ever invented.

The railway was worked not by steam engines that moved along the track pulling the carriages, but by stationary engines fixed at either end of the route. A rope was attached to the fixed engines and this was used to drag the carriages along the rail. There was one rope for the up and another for the down traffic, each rope having a total length of about eight miles and a weight of forty tons. On this line, one of the earliest electric telegraph systems was used to tell the engineer at Blackwall or Fenchurch Street when to begin to wind up or let go his rope.

On that first journey the down train, as it left Fenchurch Street, consisted of seven carriages. The two carriages at the front went through to Blackwall; the next carriage only as far as Poplar, the next to the next station and so to the seventh carriage, which was left behind at Shadwell, the first station after leaving Fenchurch Street. As the train approached Shadwell, the guard, who had to stand on a rickety platform in front of the carriage, pulled out a pin from the coupling just in time to allow the momentum of the carriage to allow it to carry on sufficiently far to ensure that it stopped in the right place.

The same process was repeated at each subsequent station, until finally the two remaining carriages ran up the terminal incline, and were brought to a halt at Blackwall Station. On the return journey the carriage at each station was attached to the rope at a fixed hour, and then the whole series were set in motion simultaneously, so that they arrived at Fenchurch Street at intervals proportional to the distance between the stations.

There were perpetual delays owing to the rope breaking and the cost of repairs and renewals was huge – so much so, that within a few years of its inauguration the system was abandoned.

STING IN THE TAIL

AMERICA, 1900

Before the advent of sophisticated electronic alarms, securing your house and property was a rather tricky business; a thousand different mechanical traps and alarms were tried over the centuries with varying degrees of success. However, once computer operated alarms became available, incredible levels of security were possible, at least for those who had enough money to pay for them. But one item has always been vulnerable to theft despite all the latest gadgetry – the bicycle.

Bicycle theft is the curse of the cycling classes – no sooner has a superlock been invented that no known cutting device can get through than the thieves hit on a solution that involves either inventing an even more sophisticated cutting device or finding a way to dig up the railing or lamppost to which the cycle is chained.

The real problem is that hi-tech modern cycles can be worth several thousand pounds and since in the very nature of their use they have to be left out in public, they are always going to be a juicy target for thieves.

Back in the day when bicycles had only recently been invented and cycling was all the rage, people were fiercely protective of their machines. Cars were still in the future and bicycles were the most exciting travel invention in a thousand years – at last men, women and children could whip along at up to ten miles an hour under their own steam and for far less than the cost of owning and running a coach and pair or even a horse and cart.

Adolf Neubauer had several bicycles stolen in turn-of-the-century New York and he decided to do something about it. An engineer with a passion for gadgets, he worked on a number of inventions that he thought would protect his beloved bike and also make his fortune.

And so was born the patent saddle spike. This looked like any ordinary leather saddle but closer inspection would have revealed a mass of tiny perforations in the top part of the saddle. Each hole housed a thick needle-like spike fitted to a steel platform that could be raised and lowered beneath the saddle. When you parked your bicycle outside a shop or wherever, you turned a small crank handle fitted on the frame just below the saddle post. This raised the steel platform with the spikes attached and pushed them up through the saddle.

The spikes stood proud of the saddle roughly three quarters of an inch and though history does not record the screams of agony of a would-be thief who happened to leap aboard Mr Neubauer's bike, it is easy to imagine that it might have happened. Neubauer apparently took a delight in deliberately leaving his spiked bike where it was likely to be taken and then keeping an eye on it from some concealed spot.

Today, of course, any bike fitted with similar spikes would cause a thief significant pain and some injury – but the result might be that the bicycle owner would be prosecuted rather than the bicycle thief.

LIFE-SAVING SUITCASE

GERMANY, 1900

In the days when long-distance journeys invariably meant at least some travel by water, a perennial worry for travellers was that their ships would sink. Despite massive improvements in the construction of ships and navigational equipment – from the timber vessels of the early nineteenth century to the iron-clad ships of later decades – accidents still happened. Commercial pressures ultimately led to the construction of a liner – the *Titanic* – whose main claim to fame, apart from her sheer size, was that she was unsinkable. That claim was designed to allay fears that had built up over previous decades – decades that had seen numerous tragedies at sea.

While the ship designers toiled to improve things on a grand scale, numerous inventors competed to come up with something for the individual that would tap into the commercial possibilities of the nervous traveller.

It was with this in mind that a German scientist – one Albert von Krenkel – came up with a remarkable, and for a time hugely popular, solution: the life-saving suitcase.

Despite its name, the suitcase was actually a large trunk. In almost every respect it looked exactly like any other large trunk, but it was made entirely of thick, highly buoyant cork and had man-sized circular detachable panels in the top and bottom.

When and if the traveller's ship ran into problems and began to sink, the lucky trunk owner simply detached the two panels, threw away the trunk's contents, stepped into the hole left by the removed panels and then pulled the contraption up until it was safely around his or her middle. The happy passenger then strolled up to the main deck before launching himself or herself on the high seas and awaiting rescue.

WATER ON THE BRAIN

ENGLAND, 1900

Those splendid pith helmets that we associate with the knobbly knees, the huge pairs of short trousers and the bad-tempered habits of old colonels in India and Africa were absolutely vital to the maintenance of Empire. Much thought and effort was expended on their manufacture because however determined the British were to bring the solid values of the Home Counties to those less well favoured parts of the world it could only be done if the poor blighters doing the work were protected adequately from the sun. Which is why one bright spark thought it would be rather good if pith helmets could carry their own water supply.

The all-purpose water helmet, advertised in a number of late-nineteenth-century magazines was built high and wide with a gentle slope down to a wide brim. At the bottom of the slope all the way round and at the edge of the brim was a deep gutter. On those rare occasions when it rained in the tropics the water would collect in the gutter and then run into a specially made miniature tank at the back of the hat and below the gutter. The tank protected the neck from the sun and could store almost a pint of water. When the wearer felt a little thirsty he took the hat off, turned it round and undid a small brass tap fitted to the tank at the back. This released the collected water that, we are reliably informed, could be used 'either to refresh the palate or to cool the brow'.

It is easy to see why the idea appealed because, in addition to providing water for drinking and washing, the water tank, being fairly thick and filled with water, kept the sun that little bit further away. However, there were drawbacks and they quickly became apparent – to such an extent, in fact, that the revolutionary new hat never became a success. The gutter tended to stay damp, which meant that fearsomely dangerous bacteria were able to live and breed there; the hat was also rather heavy and the water occasionally heated up so much that it was disgusting to drink.

NOSE JOB

ENGLAND, 1901

We are all prone to thinking that our appearance could be massively improved if only our ears didn't stick out quite so much or our nose was just a bit smaller. At the turn of the twentieth century when there was absolutely no legislation to compel manufacturers to remain 'legal, decent, honest and truthful' about their products, extravagant claims were made – and believed. And when it came to devices that claimed to improve the appearance, the poor punter was particularly vulnerable. Perhaps the most amusing invention among a host of batty cosmetic ideas was Professor Ray's Nose Adjusting Machine. We don't know if the professor really was a professor, but even if he was genuine there is no doubt that his machine was bogus – it could not possibly have done what was claimed for it.

The advertising for the adjuster was unambiguous: 'This is the only patent nose machine in the world. It will improve ugly noses of all kinds. It can be worn during sleep. Send a SAE for particulars.'

Intricate and compact, the Nose Machine had been very carefully thought out and made to exacting standards. Held in place by a latticework of linen straps that ran over, under and behind the head, the adjuster itself was made from brass plates fitted with a number of tiny adjusting screws. Once the plate had been pulled over the nose these screws enabled the wearer to tighten or loosen the machine until the nose itself had been squeezed into the required shape. The theory was that if the machine was worn every night for several months the wearer's nose would eventually retain the shape into which it had been squeezed.

Professor Ray was so confident that his nose machine was nothing to sneeze at that he patented it in several countries and – he claimed – sold more than six hundred in the first year of manufacture. We don't know how long it was before the valiant six hundred discovered their mistake, but by 1905 Professor Ray's Nose Adjusting Machine was no longer an option for the cosmetically challenged.

GLOVES GONE MAD

ENGLAND, 1902

The Victorians always covered their table legs – bare table legs for those most repressed people were almost as bad as bare human legs – but our great-great-grandparents were also rather obsessed with cleanliness. Countless children really did have soap pushed into their mouths as a cure for swearing and all kinds of improbable little inventions appeared on the scene aimed at those who would do anything to avoid a spillage or a stain.

Take gloves, for example. It is difficult to believe now but at one time no self-respecting person – male or female – would venture out of the house without a pair of gloves. Naked hands really were too provocative but when you arrived at a friend's house for tea wearing your beautiful gloves how on earth could you help yourself to a slice of cake without soiling them. The answer lay in one of the oddest bits of clothing ever invented: the glove-protecting glove. This was actually just part of a glove – it covered the thumb and forefinger only, fitting snugly over the full glove. The glove-protecting glove came in ordinary cotton or, for the gentry, it was available in a beautiful cream-coloured silk. Of course, if the glove-protecting glove was that beautiful you'd need something to protect it . . .

The glove-protecting glove was an invention of the decorous Edwardian era – it was the idea of Winifred Buckland Walley and patented in 1902. At about the same time, however, the science of clothing manufacture was faced with a far more serious problem. This might best be summed up as the great masturbation dilemma.

The Victorians and Edwardians were so obsessed with the idea that masturbation was a sure-fire ticket to the lunatic asylum that they invented all sorts of means to prevent even inadvertent masturbation, if such a thing can be said to exist.

Many inventions were a curious mix of corset and medical device. One elaborate contraption encased the female body from just above the knee to just below the breasts. It was made from leather and cotton and the idea behind it was that any woman not in full command of herself would not be able to succumb to temptation simply because the corset took nearly half an hour to remove. It was designed, according to the manufacturers, to reduce the risk of 'moral incontinence' by 'putting the wearer's body in a state of healthy uninterruptible enclosure'.

According to one later commentator, the real point of the continence corset, however, was to ensure that women with what would now be termed learning difficulties should not become a prey to sexual predators. If women didn't or couldn't say no and shout for the nearest policeman the continence suit of armour would soon make any would-be ravisher give up and head for home.

1902 also saw the introduction of the latest in a long series of moral underwear for men. Many of these devices were marketed as preventing nocturnal emissions. They were usually made in very stiff calico and with leather belts and buckles to make life difficult for wandering hands or wandering imaginations. Worried that the bar to temptation really did need to be much stronger, the moral scientists of 1902 came up with a garment that involved a polished steel plate (polished to get rid of the sharp edges!) with a downward pointing tube into which the wearer was supposed to insert his penis before retiring to bed. The steel plate was strapped to the body with plenty of complex belts and buckles (just in case there was any attempt to quickly remove the garment at night) and the wearer could be confident that any nocturnal erection would be stopped in its tracks.

GUARANTEED SOFT LANDING

AMERICA, 1902

We are all familiar with bull bars on jeeps and other off-road vehicles, but how many of us know that in the early days of the railway there were numerous attempts not just to build contraptions to push cows and buffalo out of the way, but also to catch unwary pedestrians who inadvertently strayed in front of the local express train?

At first rubber-cushioned ramps were added to the front of locomotives but experiments revealed that although they reduced the direct force of any impact they still threw the individual out of the way so violently that serious injury was the result. The greatest railway inventors of the day were called in to give their opinion and the consensus was that more cushioning was the answer. Thus the Pedestrian Catcher was born. At first catchers were to be attached to city trams in Los Angeles but after a suitable trial period the plan was to fit them to passenger trains.

The Pedestrian Catcher consisted of a long, thin, upholstered sofa mounted on small wheels and with an extra high back (the back was designed to rest against the front of the tram). The sofa catcher also sloped downwards so that it would, as it were, sweep the foolish pedestrian off his or her feet before depositing them between the comfortable arms of the higher part of the sofa.

On its maiden outing the sofa knocked the human guinea pig clean out of the way, breaking one of his ankles in the process. While officials stood around wondering whether to try again the heavens opened and within minutes the sofa was a sodden mass. The effects of the weather had been completely forgotten by those who had designed and built the catcher, but they knew when they were beaten and the sofa-fronted tram was seen no more.

FOR THE BIRDS

AMERICA, 1902

When chickens were produced at low density on farms that looked like farms the birds tended to get along quite well. Fights broke out, it is true, between rival cocks but these battles were normally settled without bloodshed. Under certain circumstances, however, chickens can be very cruel animals indeed. A genetic mutation that produces a pure white chicken, for example, will create an animal that is fairly quickly pecked to death by its fellows.

As the demand for chicken meat rose with the vast increase in world population, industrial farming processes brought new problems and some very bizarre inventions appeared in an attempt to cope with them.

The biggest problem is that if you cram a lot of chickens close together they will spend a good percentage of their time trying to peck each other's eyes out and without going back to uneconomic low-density farming there seemed no obvious way to stop this. Then a bright young scientist came up with the idea of spectacles – or, more properly, goggles – for chickens, which is not as mad as it might sound. The eyewear was designed not to enable the birds to see better and escape the peckers, but to protect those birds that were in danger of losing their eyes. The odd thing about chickens is that victims and aggressors were the same birds – while they still had their sight all the chickens would try to blind each other. The only birds that escaped the maelstrom were the ones already blinded.

The new plastic goggles were designed to slip over the chicken's head just like human goggles but where tens of thousands of chickens were involved the cost was high both in terms of having the spectacles made and the time involved in fitting them. Several men had to be employed pretty much full time running after the chickens and putting their specs on as well as monitoring them to make sure they didn't fall off. The specs, though made from a simple piece of plastic, were also expensive as they wore out and had to be made in their millions. Eventually it was decided that the cost outweighed the benefit and academic-looking chickens were seen no more. The man behind the chicken glasses? Step forward Andrew Jackson Junior, an American who patented his goggles in 1902.

WHAT TO DO WITH THE DEAD?

AMERICA, 1903

Victorian burials in British cities had their horrors and scandals. In many cases the poor were buried in churchyards already crammed with bodies, so much so that in some places gravediggers would jump up and down on a coffin to force it into the ground and then cover it with a few inches of soil. Sickened by this sort of thing, great minds tried to finds ways to deal with the dead that were dignified and respectful but that didn't pose a health hazard or take up too much space.

In the United States a far more practical view was taken of dealing with the dead and concern with churchyard burial was not nearly as central to people's way of thinking as it was in Britain – which is why various exotic methods of disposal were suggested and sometimes actually tried out. These included having one's ashes made into a garden ornament or piece of sculpture; being buried at sea (even when you'd died on land); and, more recently, being jettisoned into space.

In the early 1900s the most bizarre idea of all was invented by Joseph Karwowski. Mr Karwowski invented and patented a remarkable process for encasing your deceased loved one in a large and solid lump of pure glass.

Karwowski spent years perfecting his system, which was technically demanding and also very expensive. In order to avoid distortion in such sizeable chunks of glass he had to employ highly skilled glassmakers at huge cost and even then the results were not always satisfactory. The corpse to be encased couldn't simply be covered with molten glass or it would effectively cook and the weight of glass, if not properly handled, might distort the features of the dead person or disturb unnaturally the hang of the clothes. Moulds were made and various techniques tried out, but by the time Karwowski had produced a finished result with which he was happy he realised that there was no real market for his corpse-in-a-glass-case idea. He had assumed that people would be delighted at the prospect of their nearest and dearest being preserved from decay in glass. He thought that people would simply keep the glass blocks containing their dead relatives in the sitting room or dining room. He advertised that he would position them in the glass in a favourite attitude or on a favourite chair and must have been astonished to discover that when it was time to market the new idea there were very few takers indeed.

THE PORTABLE BATHTUB

ENGLAND, 1904

When London's Charing Cross Hotel opened at the end of the nineteenth century, church leaders were outraged. The problem wasn't that it had a built-in brothel or that it sold sex aids or rubber wear; no, the problem was far worse – the Charing Cross Hotel had too many bathrooms.

At a time when even the biggest hotel would have had at most half a dozen bathrooms the Charing Cross Hotel had almost one bathroom for every two bedrooms and the bishops assumed that if hotel guests were that keen to be clean it must be because they hoped to go bed hopping morning, noon and night. Cleanliness it seems is not next to godliness where the temptations of pleasure are involved. In some parts of the world cleanliness can be a big problem. Intrepid travellers who find themselves in waterless hotels in remote regions of the earth have always longed to plunge into a refreshing tub filled with warm water. With that in mind a number of inventors have come up with portable or traveller's baths. All have one thing in common: they are completely mad.

Perhaps the silliest is the bag bath invented by the German scientist Adolf Herz in 1904. This was a rubberised bag roughly six feet long and four feet wide with a hole for your head and a zip up the front. The idea was that you climbed inside the bag, zipped it up and then found a convenient hose pipe or bucket to fill it. Once you felt there was enough water in the bag the instructions suggested you should jump up and down to make the water slush around you with enough violence 'to effect a crisp cleansing action'. For a really thorough wash ten minutes' jumping was recommended. For a refreshing but not necessarily cleansing bag bath two minutes' light hopping would do the trick.

As late as 1972 the portable bath idea still gripped the imaginations of numerous inventors and Francis Allan's modification of the basic Herz bag bath included a valve to allow the user to safely drain away the water after bathing. The Herz original could only be drained by tipping it over, which meant hurling several gallons of water out of your hotel room or flooding it!

A Japanese bath bag for use in jungle settings or deserts involved a similar bag but this time it had holes cut for arms and legs. The holes were fitted with elasticated grips to make them waterproof and the wearer could therefore take a bath and travel at the same time!

EDIBLE MUSIC

AMERICA, 1905

The fact that something is difficult to do is not necessarily a good reason for doing it. But inventors don't often see it that way, which may explain why there are many examples of scientists and inventors spending half a lifetime (not to mention a small fortune) trying to produce something that the rest of us think of as at best superfluous and at worst downright crazy. Take edible knickers, for example, or singing lavatories or – perhaps best of all – trousers fitted with an electrical heater.

Novelty inventions are slightly different because the main appeal here is making money: an apparently silly invention can catch on and lead to huge sales precisely because it is a bit silly. Left-handed mugs were all the rage in the 1970s, for example, and lava lamps have long been popular for reasons that baffle most of us. One or two inventions are so crazy that they seem to blur the line between novelty and serious invention.

The record player designed to play chocolate discs is probably as good an example as any. Why on earth, you might ask, would someone who buys a record player want to buy records that have to be eaten? Surely a music lover would not want to eat (and therefore destroy) a record he or she enjoys listening to. No such apparently logical consideration clouded the mind of John William Mackenzie who, in the first decade of the twentieth century, proudly patented his phonograph with phonograph plates of edible material.

The seriousness with which Mackenzie viewed his creation

can be judged from this extract from his patent application:

> The object of this appliance is a phonograph, which embodies the principle of the phonograph in an easily understood manner. A special feature of this phonograph is the plate or the bearer of the sounding record, which is made of some edible material, more especially of chocolate. This material, which has up till now been unknown for such a purpose, is specially well adapted for the plate or cylinder of a phonograph and is well adapted to make, as it were, sweetmeats speak.
>
> Chocolate mass is most suitable for this purpose, as in a sufficiently warm state, it takes on exactly the record of the phonogram by impression of the negative and retains same in true reproduction on becoming cool. But also other formable sugar masses, which are used in confectionary for making fruit, figures etc. can be used for the purpose.
>
> Chocolate is specially adapted for phonogram reproductions. Edible material of this kind can be rendered more suitable for the manufacture of edible phonogram reproductions by covering same with metal foil, for instance tinfoil.
>
> The tinfoil or similar covering as well as the edible material thus becomes the bearer of the phonogram. The record thereby becomes more durable and outside noises that arise from the movement of the pin over the hard chocolate or confectionary are easier avoided when the record is phonographically reproduced.
>
> Apart from the useful purpose of the tinfoil covering for the phonographic effect, the edible material, for instance the speaking chocolate plates or cylinders, thereby remains more appetizing for eating. Metal foil as covering for the edible phonogram bearer can also be substituted by some other pliable substances.

Did the inventor want the chocolate discs to last or to be eaten? Was he interested in eating or music? One thing is for sure: the edible record player did not take the world by storm.

CART BEFORE THE HORSE

AMERICA, 1905

The Edwardian era both in Britain and America was a time for throwing off the shackles of the long, dull nineteenth century and embracing change – early motor cars were seen more often, experiments were being made in powered flight, trains had reached undreamed-of heights of efficient luxury, and science and technology generally had seemed to open up a brave new world.

Central to transport was still the horse, however, and a number of inventors were convinced that with a little ingenuity horse transport could be dragged into the modern technological world.

The splendidly named Serge Berditschewsky decided that horse transport could be revolutionised in such a way that it would be just as good (if not better) than this new, noisy, slow, inefficient upstart the motor car.

Berditschewsky came up with an odd-looking carriage whose unique feature was that it completely concealed the horse. Berditschewsky was convinced that the real objection to horse-drawn vehicles was that the driver and passengers had to look over the horse to see ahead. The Berditschewsky carriage put the horse behind the carriage and inside a completely enclosed compartment. Within this compartment the horse was fitted with a special harness that centred on a wide breast belt against which the horse was trained to push. When it pushed against the leather belt the energy was transferred to the wheels of the vehicle, which would continue to move so long as the horse continued to push.

An elaborate system of pulleys led the reins back from the front of the carriage into the horse's compartment and the driver had to pull a special lever if the horse needed a little encouragement – the lever operated a riding crop that gave the poor horse a smart thwack on the rear end. Pulling on the reins slowed the horse down in the usual manner. Berditschewsky was convinced that since his new carriage, though rather large, looked as if it had dispensed with the services of the horse, it would be as good as primitive motor cars that really had dispensed with the horse.

Little did he know that cars would soon make his and every other horse-drawn or -pushed vehicle redundant in the developed world.

IT'S A DOG'S LIFE

ENGLAND, 1905

The Englishman's love for his dogs is legendary, which may explain why the Victorian Earl of Pevensey had himself buried with his favourite spaniel, and why Lord Derby dined every day with his ten dogs seated round the table, each served in turn by a team of flunkeys.

The crafty German scientist and entrepreneur Johannes Rohr decided to cash in on this passion for all things canine by designing what he described in his first printed advertisement as an Improved Humane Kennel for Man's Best Friend. The benefit of the kennel – which was to be produced in a range of designs from miniature Gothic castle to suburban villa – was that, using a Heath Robinson arrangement of weights and counterweights, it would keep the dog warm and cosy at all times.

As the dog entered the kennel, its weight, as it stepped on a platform, operated a pulley that, in turn, released a catch that lowered a miniature duvet over the lucky animal. When the dog walked out of its kennel the system went into reverse and the duvet was lifted back up to the top of the kennel.

At first the difficulty was that dogs tended to bolt in terror as the cover was let down, or they tried to eat the duvet; one poor animal managed to get tangled up in the pulley system. Hardly had the Improved Humane Kennel begun to take the pet world by storm when it was suddenly – and permanently – discontinued.

LEAP-FROGGING TRAINS

AMERICA, 1905

The early railway was condemned by the Church. Bishops fulminated against the blasphemy of travelling at fifteen miles an hour and predicted an increasing number of railway disasters as God showed his displeasure at the new sacrilegious form of transport. Doctors predicted dreadful internal injuries as speeds increased and internal organs were increasingly jostled about. Certainly the very early railway was dangerous: railway lines were not fenced off; amateurs sometimes put their private locomotives on the tracks and set off in any direction they fancied without warning the authorities and sometimes continuing until they hit an oncoming train that was part of a scheduled service. The difficulty was making the law and the regulations keep pace with the technology.

Great minds on both sides of the Atlantic wrestled with the problem of how to reduce the risk of accidents in an age when super-efficient engines and brakes were simply not available. Their efforts threw up some wonderfully bizarre inventions and few were more bizarre than the anti-collision system dreamed up by the American railway engineer R K Stern. Mr Stern had started out as an electrical engineer but he turned his mind to a number of railway engineering problems when he realised that railways were booming.

Stern's great idea was to solve the problem of head-on collisions where there was what railway people called 'single-line working'. Single track was common throughout the system in Britain and America until the end of the twentieth century and beyond, and numerous methods were tried to ensure that there was absolutely no way that two trains could end up travelling towards each other on the same track. It might sound unlikely that they should ever do so but head-on collisions on single track were relatively common. The solution was eventually found in England, where a system was instigated by which the driver of a train had to have a baton before he could proceed on to a single line. Since there was only ever one baton for each piece of line there was never a danger of two trains colliding.

Before the baton, Mr Stern decided on a much more complex answer to this vexed issue. He built his train with a sloping front and rear, and running up each of these slopes and then over the top of the train was a section of track that exactly matched the track on which the trains were running. In theory, if two trains found they were on a collision course they needn't worry about braking because instead of smashing each other to pieces at the point of impact, one would simply run up over the top of the other and then down back on to the track; a sort of leap-frogging that would put a permanent end to head-on collisions.

Mr Stern built a prototype of what he called his leap-frog railway at Coney Island Pleasure Grounds in New York in 1905. He hoped that if it could be shown to work here it would be adopted on real passenger and freight lines all over America. The leap-frog railway was by all accounts a huge success – travelling at 8 to 10 miles an hour, the leap-frogging trains shot over the top of each other with no problems although passengers confessed they were terrified. Despite the success of this small-scale effort, though, the principle was never adopted on the national network, which was probably just as well for differences in mass and weight and energy would almost certainly have led to disaster. So that fairground attraction remained just that – a fairground attraction – and Mr Stern went on to other things.

FLY SWATTER

GERMANY, 1906

From whisks made of horse hair (popular with African dictators) to all kinds of strange electrical devices, fly swatters come in a huge range of styles. In some countries, of course, fly swatting is more important than in others – successfully swat your fly in Africa, for example, and you may avoid malaria or worse. That may explain why numerous inventors have turned their attentions to the task of creating the ultimate fly swatter.

One early twentieth-century German invention involved a clockwork device attached to the end of a stick. You wound up the device, which looked rather like a miniature version of the paddles on a paddle steamer. Then when your fly buzzed into view you pressed a catch on the stick and the paddles whirred violently round. You then chased the fly with the stick while the paddles were still whirring and with luck the paddles battered the fly to bits. There was one drawback with this invention apart from the fact that it was beautifully made from brass and mahogany and therefore very expensive. The problem – a critical one – was that it didn't work.

An early-twentieth-century invention tried to go one better by catching the fly in a net. The idea was that using a spring-loaded rod you could fire a net about one foot in diameter and catch half a dozen flies at once. The fly net was, like the paddle swatter, beautifully made, with a silken net with a fine brass rod. But once again there was an insurmountable difficulty – it didn't work.

An invention that did work was the disc-firing fly-swatter gun. Invented in America in the 1950s, it was a simple spring-loaded handgun (rather like an air pistol), which fired a flat metal disc. You had to attack your fly while it was sitting on a wall or curtain or the device usually failed to work. The drawback of not being able to dispatch your bluebottle in mid-flight was that you ended up with unsightly splatters of blood and bits of fly all over your walls.

Far better was the electro-stun fly swatter, again invented in America. This was expensive but emitted a strong electric pulse that stunned any fly within a two-foot radius of its coil. The drawback? Children got hold of it and tended to use it on each other!

But perhaps the most eccentric fly-swatting invention of all was George Blake's 1923 suction cup gun. Big, beefy and completely out of proportion to the job it was supposed to do, the Blake suction gun had to be cranked up like a medieval crossbow and was as likely to drive a huge hole in your wall as kill the odd bluebottle!

ALARM BELLS

AMERICA, 1908

Pest control has always been a fertile area for the creative energies of scientists and inventors – hardly surprising given the huge damage pests could cause in an era before chemical and electronic aids. In the worst instances, rats and mice particularly could devastate vast areas of crops or make houses completely uninhabitable. Anyone who could come up with a cheap, effective solution was sure to find him or herself rich and famous.

One bright Victorian spark devised a rat and mouse trap that lured unsuspecting rodents on to the top of a small barrel; there they were faced with a piece of hard cheese or other tough edible substance that required a great deal of tugging if it was to be removed or eaten. The movement imparted to the barrel top by the tugging – in theory at least – released a powerful spring and the small platform on which the mouse or rat had been standing was propelled upwards with enormous force. The concept was that the mouse or rat would be killed by the sheer force of acceleration or, failing that, by high-speed contact with the kitchen (or other) ceiling. The problem with this trap was that the person setting it was liable to be hit in the face by the barrel lid while trying to prime the device, or the mouse would spend an hour tugging at the cheese without triggering the mechanism. Worst of all, on those rare occasions when the trap did work, the mouse or rat would be splattered over the ceiling and bits of it then had to be picked up from various parts of the kitchen.

A more reliable if equally batty idea was produced by Joseph Barrard and Edward Markoff in 1908. These stalwarts of the pest control business spent years working on the whole theory of pest control. What was the point they asked in simply killing individual pests when what was really needed was a way to prevent rats (they specialised in rat infestations) causing a problem in the first place. In other words, they believed in prevention rather than cure. To this end they decided that their trap would use one rodent to scare many rodents. They created an elaborate device which, when baited, attracted any passing rat into a narrow tunnel where it was rewarded with a piece of chocolate or cheese, but in order to get its reward the rat had to put its head through a special device. This was fitted with a small elasticated collar that remained on the rat's neck when it withdrew its head. On the collar was a bell. The idea was that the rat would panic on finding itself with a collar and it would run frantically among all its fellow rats. The noise of the bell would terrify the other rats, which would then run away from the bell-ringing rat, but as fast as they ran away (from the crops, the house or whatever) the faster the bell-ringer (seeking comfort among its fellows) would run after them. The end result would be that huge packs of rats would start running and never stop.

Of course, in practice the rats managed to avoid the collar and on the few occasions when it did stay on the rat's neck the other rats quickly got used to it and realised it was harmless; for Barrard and Markoff it was back to the drawing board.

NEW BROOMS

AMERICA, 1909

When vacuum cleaners first arrived on the scene they were one of the wonders of the age – contrary to popular belief, the earliest vacuum cleaners are Victorian. They were not terribly efficient because without electricity the suction effect of a reasonably powerful vacuum could not easily be achieved. All sorts of bizarre pumps were created to try to generate a vacuum effect and some were reasonably successful though nothing like as powerful as their modern counterparts.

An early American example used two people – one to guide the vacuum cleaner around the floor, the other to move a large lever back and forth above the box on wheels that held the dust bag.

Then in 1909 an American patented what must be the most eccentric vacuum cleaner in history: the Behringer Vacuum Cleaner. The main part of the cleaner was built into the base of an upholstered mahogany rocking chair. To all intents and purposes this really was just a chair and only a close observation of the bottom part would reveal that it housed an elaborate device to create a vacuum.

In the days when men expected women to do all the housework, the chair was also well ahead of its time – the manufacturers recommended that the wife should push the sweep part of the vacuum cleaner around the floor while her husband sat in the chair rocking violently back and forth to create the bellows effect that would – in theory at least – suck up all the dirt.

It is extraordinary with this as with so many strange inventions that no one thought of what now seem like obvious drawbacks. How on earth, for example, was the chair to be carted up and down stairs and around the house every time a different room had to be cleaned? And what would happen if the husband worked late every night and at weekends or the owner of the cleaner lived alone. So far as we know there was no warning on the box that the armchair vacuum cleaner should not be bought by single people.

FULL STRETCH

ENGLAND, 1910

The virtuoso composer Franz Liszt made piano playing into a sexually charged activity on a par with the cavortings of Mick Jagger or the Gallagher brothers. Women were warned not to attend Liszt's concerts on the grounds that they were improper – 'The man is a sewer!' declared one bishop – but the swooning of young (and sometimes not so young) women meant that Liszt enjoyed the favours of a string of musical groupies long before the word had even been invented.

Some of the attraction had to do with the passion with which Liszt played but it was also down to the fearsome difficulty of the pieces he himself composed. Much of the difficulty had to do with the stretch required to play some of his chords. For those with even relatively small hands an octave stretch on a piano can be tricky but Liszt sometimes demanded an octave and a quarter or more, which explains why, by 1910, the inventors had come up with a fearsome-looking device called the finger stretcher.

The advertisements for this medieval instrument of torture claimed that 'musical science' had spent decades perfecting it and it is certainly true that the stretcher, made in brass and mahogany, is a precision instrument and an attractive one at that. However, it beggars belief that those musical scientists could really believe that their invention would work. Certainly they were exploiting the wild enthusiasm for playing like Liszt

– and to play like Liszt it was necessary to be able to play those demanding chords.

The finger stretcher worked in the following manner: the little finger was placed in a sliding wooden block fitted on a brass rail that was itself embedded in a length of mahogany. At the other end of the mahogany plank was another sliding block of mahogany into which the thumb was placed. The two blocks were then forced along the rail in opposite directions to stretch the distance between little finger and thumb. Once the blocks had been pushed as far apart as possible with the pianist's thumb at one end and little finger at the other, the two blocks were locked in position. In theory, after a few days that amount of stretch would increase and the blocks would gradually be eased further and further apart until the pianist was able to play those impressive Liszt chords and presumably enjoy the favours of all those fainting women.

ALCOHOLIC REVIVER

ENGLAND, 1910

Edgar Stanhope, an Oxford scientist who also happened to be a keen angler, carried out a number of experiments using brandy as a means of restoring life to dying fish.

Having kept a trout out of water until it had apparently died he would then drop it into a bucket filled with undiluted brandy. On the first occasion he tried this he commented: 'It was highly interesting to see the plucky manner a trout battled with his fainting condition, after a dose of brandy, and came out the conqueror.'

On his next visit to the Wye, Dr Stanhope took his bucket of brandy out in the boat – to the amusement of the local gillies – and tried the same experiment with a salmon. The results were less impressive, as Stanhope himself admitted: 'Strange to say, the salmon did not once attempt to rouse himself after being dosed, the consequence being fatal to him. This was the only fish that succumbed under the treatment.'

John Bickerdyke, an author and amateur scientist, then tried the experiment with a few coarse fish. He was most impressed by the effect of brandy on the dace.

'I had him out of the water three times of five minutes each. He was exceedingly faint and almost dead, but immediately the brandy was given he pulled himself together and in the course of a few minutes not only recovered, but darted around with a rapidity positively amazing.'

DON'T BE ALARMED

AMERICA, 1912

One of the challenges of inventing has always been this: how can you make an effective burglar alarm for people who are deaf? The obvious answer – flashing lights – was clearly far too simple for at least one early twentieth-century brain.

In 1912 Arnold Zukor decided that he had the solution and it was a far more interesting and exciting one than the rather mundane business of setting up a series of popping lightbulbs.

He invented a sprinkler system rather like the ones now fitted to most offices and shops. But where a modern sprinkler system is triggered by smoke, Zukor's system was set off by an external window being lifted or broken or an external door being opened. The alarm system was expensive and involved pipes being buried in the plaster work all over the house and running back to a main supply used to feed water to the domestic house system. A high-pressure pump ensured that when the system was triggered it didn't just rain gently on the occupants of the house but came out of walls and ceilings in torrents.

Zukor doesn't seem to have considered that apart from the sheer costs of installation his system had one or two other major drawbacks – for a start the effects of so much water pouring on to furniture and fabrics would be as financially damaging as having them stolen.

But Zukor worked on the principle that the worst thing in the world was that someone else should damage your things – better that they should be destroyed in front of your eyes.

SYNCHRONISED PROPELLER

GERMANY, 1918

When aeroplanes were first used in warfare during the 1914–18 conflict they had a major drawback. Bombs could be thrown out by pilot or observer (as the second man was then called) because they could be lobbed out sideways before disappearing into the slipstream and down on to the heads of the enemy. But for a fighter to be really effective the pilot had to be able to fire in the direction in which he was travelling, but that meant firing a machine gun through the propeller, which of course on early planes was mounted immediately in front of the pilot. Early experimenters thought the bullets would travel so fast that they would not hit the spinning propeller; in practice they smashed it to pieces. What was to be done?

Roland Garros, a Frenchman, tried to solve the problem by covering the early wooden propellers with steel – this worked after a fashion but pilots were always in danger of being caught by a ricochet and, besides, it was wasteful and inefficient with half the bullets clattering off the propeller in all directions.

The German aircraft engineer Anthony Fokker came up with a better idea: his ingenious answer was to create a synchronised machine gun that shot through the blades by firing precisely in sequence with the propeller itself – the bullets left the gun only when there was a gap in the rotating propeller in front.

That was a relatively sensible invention but it was based on a much madder idea that had been tried a decade or so earlier.

An edition of the *Strand* magazine for 1901 mentions a device issued experimentally to Army snipers. This consisted of a wind-up whirring fan with metal blades. When the sniper wanted to peep out from cover to see what the enemy was up to he would wind up his bullet-stopping fan until it was going at a terrific speed and then, placing it in front of his head and chest, he stepped out. He could see through the whirring blades but if an enemy sniper shot at him the theory was that the enemy bullet would be harmlessly deflected by the whirring blades. When the sniper with the fan wanted to shoot he would momentarily stop the blades whirring, shoot quickly and then start them up again.

HEAD SHOT

AMERICA, 1918

Albert Pratt deserves to be better remembered. We know very little about him, but Mr Pratt – an American – was a passionate inventor who came up with dozens of devices to help defeat the Germans in the First World War.

One of the most outlandish of these was the gun helmet. A conical hat with peaks fore and aft, the gun helmet had a large-bore barrel protruding from the front with a metal sighting device descending from it. Roughly six inches long, the barrel was rifled (grooved internally) to increase accuracy and it was fired by an ingenious air-pipe that doubled as a chinstrap. When the soldier wearing the hat wanted to fire he blew into his chinstrap; the air travelled up the chinstrap (a narrow, flexible pipe) and inflated a rubber bulb inside the hat. As this expanded, it pressed on a trigger that fired the gun. The escaping gases from the gunpowder were used to re-cock the gun for the next shot. The great advantage of the shooting hat was that it allowed the soldier to fire from the head and, simultaneously, from the hip with his more conventional rifle. Better still, the hat didn't encumber the soldier as a conventional second firearm would have done. What's more, it was fairly accurate – the soldier aimed the sighting device (a bit like a short metal calibrated ruler dangling from the front peak of the hat) until a particular mark was lined up on the target and then gave a quick puff down the tube; in theory at least, another enemy soldier was immediately out of action.

However, the shooting hat is a perfect example of how inventors sometimes let an idea run away with them without looking at some of the more obvious drawbacks.

As soon as the shooting hat was tried on a few soldiers it was realised that the noise of each shot, combined with the recoil, quickly gave the soldiers terrible headaches as well as making them dizzy and, after a while, completely dis-orientated. For Albert Pratt it was time to think again.

EYE MASSAGER

AMERICA, 1920

For commercial companies the Victorian era was a godsend. Technology had vastly increased the potential for manufacture on a large scale and a relatively innocent public had not yet been protected by laws restricting what manufacturers could say about the things they made – which explains why one American company was able to market a beautifully crafted and very expensive device that massaged the eyes.

It sounds straightforward enough and something we all might enjoy, except for the fact that this particular 'massage' simply involved puffing cold air on to the eyes. Why that should differ in any way from the effects of keeping one's eyes open on a blustery day is hard to tell, but the eye massager certainly looked the part.

You have to imagine a pair of binoculars fitted with eyepieces at one end and rubber bulbs at the other. The two eyepieces of the binoculars are attached to each other by hinged brass rods that allow height and width to be adjusted very precisely. The elaborate brass rods and hinges converge on a mahogany handle.

The massager was used in the following way: you held the wooden handle in one hand and adjusted the nuts and screws on the brass hinging mechanism until the eyepieces were just where you wanted them; then, while holding the device as steady as possible, you squeezed first one rubber bulb, then the other. Each squeeze caused a little puff of air to hit you in the eye. And that was it – all that elaborate and carefully fashioned brass and mahogany simply puffed ordinary air into your ordinary eyes. The eye massager was expensive too – in today's terms it would have cost over £100; truly for the man or woman who has everything.

ROLLS-ROYCE SPADE

ENGLAND, 1921

It's been said that science tends to appeal to those who are in some way obsessive – thus evolutionary biologists will happily spend decades studying nematode worms or tiny parasitic mites that live exclusively on the claws of hermit crabs. To the rest of us this seems batty in the extreme but to experts in biology any interesting results the research throws up may conceivably have relevance in medicine or genetics or human biology.

But what are we to make of science that invents things with an immediate practical aspect that – despite being practical – is still apparently pointless?

In 1921 one William de Camp decided that the humble garden spade could be massively improved if it took some responsibility for what it was doing – after months of work he decided the best thing to do would be to fit his new spade with a complex counting device. This was operated by pressure on the edge of the blade – the top edge, that is, where the digger applies pressure with his or her foot to drive the spade into the ground. Of course sometimes the spade would be used in such a way that the foot would not be employed to shunt the thing into soil or cement – in this instance the extra pressure on the blade caused as it lifted its load would be enough to operate the counting mechanism.

This was to be the ultimate – the Rolls-Royce – of spades. But it was a delicate instrument with its various wires leading up inside the shaft to the meter at the top of the handle and early examples had a tendency to stop working after a hard day's use. There was also the difficulty of persuading builder's merchants and garden tool suppliers that the extra cost of the fancy spade was justified by improvements in productivity and usefulness.

Mr De Camp, naturally, had no doubts – he once said that the idea behind the spade was to prevent those mixing cement forgetting how many spadefuls of dried cement they had added to the wet mix and potentially ruining it.

A classic example, perhaps, of using a sledgehammer to crack a nut!

LIPS TO DIE FOR

AMERICA, 1923

The past decade has seen some extraordinary new approaches to cosmetics, with botulin poison injected into people's faces to remove wrinkles and a seemingly endless search for the ultimate herb or plant extract that really will get rid of lines of cellulite, bags under the eyes, sagging bosoms and so on. The desire to remain youthful-looking has reached almost epidemic proportions too, with plastic surgery and face-lifts, tummy tucks and bottom implants almost commonplace in Western countries.

Central to a beautiful face is a fine pair of lips, which is why so many celebrities – particularly minor ones – have gel implants in their lips. But there is nothing new about the desire for a full, sensuous, pouting mouth. Back in 1923 an American inventor came up with an idea for a machine that would produce luscious lips in a natural pout.

The lip compressor was fitted over the mouth and strapped to the head. It consisted of a brass circular clamp fitted with pads that could be adjusted to fit any mouth. The lips were inserted between the clamps each evening before bed and the screws tightened until the requisite pout was achieved. In the morning the newly trained lips were said to retain something of the shape imposed on them during the night. The manufacturers admitted that the pout would quickly disappear after the compressor was removed but they insisted that repeated use over a number of years would gradually lessen the extent to which the lips returned to their normal state.

The compressor was also recommended for short-term use – the lips, we are told, should be left in the compressor for an hour before a ball or important party and removed just before dressing. Enough pout would remain for the pouter to be made the belle of the ball.

TELEPHONING THE DEAD

AMERICA, 1925

The great inventor Thomas Edison whose telephone was and is arguably one of the most important of all inventions, was also occasionally tempted into scientific adventures of a most eccentric nature.

In the 1920s he tried to invent a spirit communicator – a device to enable the living to get in direct touch with the dead. He spent years on the project but – perhaps unsurprisingly – reported no success at all.

When he died a well-known clairvoyant called Sigrum Seuterman claimed that he'd heard from Edison and the great man was still at it, trying to make his spirit communicator, only now he was at work on the other side!

The science of getting in touch with the dead via modern electronic gadgetry is called necrophony with the emphasis very definitely on the second syllable. It all began with the invention of the telephone, but tape recorders were another boon to mediums and others. Why the tape recorder should be better at hearing the voices of the dead than living people's ears, no one knows.

A passionate believer in the afterlife and the chatter of the long dead, Constantine Raudive collected more than 70,000 tapes, each purporting to contain voices from beyond the grave. And, like Edison, he just kept at it after he died – communicating regularly with other mediums still in the land of the living. He explained that he was continuing his work of taping the voices of the spirits. His spirit voice was also recorded by his still living followers.

A device called a spiricom, which was basically a complicated radio transmitter and receiver, was claimed to have massively improved the quality of communications from the dead. There was widespread confusion and embarrassment, however, when the inventor of the spiricom sent a message after he himself had died explaining that the dead communicated on 68 megahertz. The spiricom on earth had been operating at 29 megahertz, which meant that the so-called messages it had been receiving from the dead could not have come from the dead at all – unless the spirit of William O'Neill was getting in a muddle. This prompted long debates about whether spirits could make mistakes just like humans or were infallible in these practical matters.

The advent of computers and their massively enhanced communications capacity has given another boost to those trapped in the spirit world. Their friends on earth confidently predict that blue tooth and wireless technology will soon have distant signals being picked up and recorded all over the place. Cynics believe that, like so many earlier technologies, computers are unlikely to give the dead a completely convincing voice in world affairs.

Several American companies are on the brink of marketing the sort of software that will work effectively when it comes to communicating with the dead. Edison would have been delighted and, wherever he is now, perhaps he is smiling gleefully that Microsoft may be able to do what the spirit communicator could not.

A REAL PAGE-TURNER

ENGLAND, 1927

Trivial inventions are among the most endearing – like the cup with a battery-operated bottom. Hold the cup to your lips and the bottom gradually rises, pushing the liquid up and into your mouth. Or take the cigarette holder fitted with a thermometer – as the smoke travels down the barrel of the holder, it registers the temperature. The device was marketed – during the Roaring Twenties – as perfect for serious smokers who wanted to keep a cigarette going until it reached a certain temperature (the cooler the smoke, the more satisfying it was) but were reluctant to guess. These serious smokers needed to know precisely when to throw that butt away.

But the ultimate in trivial inventions has to be the automatic book-page turner. A miracle of sophisticated technology, the electronic page-turner consisted of a plastic tray with raised adjustable edges. The book was placed within the tray and the edges moved until they just about prevented the pages from flipping over. When you'd read your page you pressed a small button on the edge of the tray. This activated a roller which pushed one page out from under the holding edge of the tray. You then reached forward and pushed the page the rest of the way across to the other side. Expensive and really quite carefully engineered, the book-page turner might have been a commercial success – if only as a novelty – if it had actually done all the turning for the reader. In fact all it did was to release a page at a time for the reader to turn himself.

The inventor might have been relying on the fact that there were millions of readers quite happy to do most of each necessary page-turn but who would nonetheless prefer not to have to hold the book while reading. But if that was the theory, it failed miserably and the automatic book-page turner was seen no more.

FLYING SUBMARINE

AMERICA, 1928

The arms race gives inventors unique opportunities. Few countries can resist an idea that may save lives or money or help them win a war or control their populations. No sooner has someone designed a new shell that will pierce the latest tank armour than another company (sometimes the same one) develops a new form of defence that will defeat the new form of attack, and this process continues indefinitely. The interesting thing about it all is that the arms companies themselves don't want to produce something completely unbeatable because they would slow down the rate at which tit-for-tat products can be produced, marketed and sold, and selling products is what the arms equipment companies really want to do.

Anti-personnel mines – designed to blow a leg or a hand off – are produced by the million in many parts of the world and even apparently civilised countries such as America have refused to become signatories to an international agreement to stop making them. The Americans (and Chinese) are presumably convinced that none of their own children will ever stumble across one of these truly absurd inventions. But of course no sooner has an anti-personnel mine been designed and sold than the company that made it will offer an expensive invention designed to track and destroy harmlessly that new anti-personnel mine.

In a sense, all military inventions are mad, but even within the mad category there are degrees of insanity and at the extreme end of lunatic invention is the amphibious military aeroplane. Now, technically the idea is well within the capabilities of modern industrial design. Float planes land on and take off from water in their tens of thousands all over the world every day of the year and once a plane has landed it might well be useful to batten down the hatches and set off, to all intents and purposes like a boat. Wings could be made to fold inward, a propeller could be fitted and a conning tower (rather like that on a submarine) constructed. Such an idea has been put forward on a number of occasions but the costs of making this hybrid have always seemed to outweigh the benefits.

What you actually get for your money is a not very efficient boat and a not very efficient aeroplane.

What must the Admiralty have thought when they were offered an already patented aeroplane that doubled not just as an ordinary boat but as a submarine, too? The patent, though detailed and logical, would have required engineering effort on a huge and cripplingly expensive scale. The Admiralty wisely decided that the vast cost was not outweighed by the benefits. Indeed it was very difficult to see what possible benefits there could be, which is perhaps odd because on the face of it one might be forgiven for thinking that a plane that could land and turn itself into a submersible would give one country a definite military advantage over another. But that's the initial reaction; when the military experts looked at the plan in detail they found it was more a novelty idea – perhaps for the Army that already has everything – rather than a truly practical proposition.

FRUIT GUARD

AMERICA, 1930

At one time or another everyone has spilled something down their favourite shirt or trousers. The resultant rush to the loo to pour water over the offending stain or the bizarre practice of tipping white wine all over it followed by salt can actually turn a crisis into a major rage-inducing drama (since the stain usually gets much worse anyway).

A slightly obsessive American inventor got really fed up with staining his clothes while eating and decided to do something about it. Despite any amount of red wine, toothpaste, and beefsteak gravy his real horror was spurting grapefruit juice. He explained that he found grapefruit juice particularly annoying because it not only stained his clothes but also had a tendency to hit him in the eye.

He was, by all accounts, advised to solve the problem simply by changing the way he ate his grapefruit – instead of presenting it to him in the usual way, with a spoon, his wife tried chopping it into small pieces for him. He was apparently furious and exclaimed, 'I do not intend to be defeated by fruit – I want to eat it as I've always eaten it and I will!'

Several months later he had come up with a device with which he was so pleased that he decided to patent it. He refused to accept any arguments to the effect that there were likely to be very few fruit eaters as concerned as himself about the lamentable tendency of grapefruit to fight back – he was convinced that grapefruit eaters all over the world would flock to stores selling his grapefruit shield.

This consisted of a curved metal lip that fitted on the breakfast bowl around the curved side nearest the eater. Once in position it effectively created a shield about one inch high and running almost halfway round the bowl. When the spoon was pushed on to the fruit any violent spray hit the shield instead of the poor devil wielding the spoon. The grapefruit shield inventor's name was Joseph Falleck and he died if not a disappointed man then at least rather bewildered. Like so many inventors he was convinced his grapefruit shield was a major contribution to the general happiness of nations so it saddened and confused him that it never caught on.

STATIC

AMERICA, 1931

The Van de Graaff generator produces static electricity for no real practical reason – basically it produces masses of the stuff we all experience now and then when we shake someone's hand or touch a railing and feel a slight shock. What's happening is actually very complicated and to do with the atomic structure of different materials but in essence it involves one material capturing electrons from another material – it's that act of capture you feel when you get a static charge.

Robert Jemison Van de Graaff, an American physicist, invented the Van de Graaff generator in 1931; some versions of the device, which was designed to smash atoms, can generate voltages as high as 20 million volts. So the Van de Graaff generator does have a practical application but where the unexpected element of the story enters is with the invention – and again it was a serious, fully patented invention – of a hairstyle based on the Van der Graaff generator.

For years small Van der Graaff generators – which consist of a smooth metal globe on some sort of a stand – were set up in amusements parks and fun fairs to entertain children. If you place your hands on the ball and turn the machine on, the static immediately runs through you, making your hair stand fiercely on end. For a while there was a craze in America for having your hair cut while it was standing on end in this way. It ensured that every hair on your head was cut to exactly the same length.

Then a clever hairdresser decided it would be even more popular if he cut the hair while it was standing on end and then sprayed it with varnish so that, once the generator was turned off, the hair stayed up on end. And thus was the Van de Graaff haircut invented!

BIRD BEATER

ENGLAND, 1932

Stanley Duncan was one of the best-known wildfowlers of the twentieth century, although his long career began in the nineteenth. He shot geese and duck all along the Wash and the Norfolk coast and elsewhere, and became well known through his letters and articles in various magazines and newspapers.

He was also a talented, imaginative inventor who was always trying out ways to improve his guns and the mechanical calls used to lure birds to him. His most extraordinary idea was to use an aeroplane as a beater.

When October came, the experiment began. A field in Yorkshire where geese were known to feed regularly was left undisturbed for a few weeks to allow the birds to settle in. The geese that came in were watched continually and as many as 24 at a time were discovered. Duncan decided this was sufficient for his friend's experiment with the aeroplane. The idea was that the pilot would take off and then fly low over the hedge at the edge of the field where the geese were feeding. The plane, it was thought, would put the birds up, but would also drive them forward toward two guns waiting at the far end of the field.

The first attempt failed miserably. The plane buzzed over the hedge but the geese had clearly heard its engine long before it reached them and they flew almost vertically, high into the sky before swinging quickly away to the left long before the plane even arrived.

The pilot landed again. The two guns got into position on the off-chance that the birds would return. Two hours later some sixteen geese alighted on the field and began to feed. The wind had risen and this may have contributed to the fact that the birds, on the second attempt, did not hear the plane's engine until it was too late. The plane rushed over the hedge about sixty feet above the geese. The birds were apparently so astonished that they made no attempt to fly off and flattened themselves against the ground instead. One of the gunners fired and the geese got up and soared away downwind, but the presence of the plane kept them close to the ground. Some tumbled into a hedge bottom, others managed to gain height before the airman, turning quickly, got above the flock again, pushing the geese towards the guns. At the end of the evening most of the flock had been bagged and the experiment was counted a success. For some reason, perhaps because it gave the birds so little chance, it was never repeated.

A PAT ON THE BACK

AMERICA, 1936

Inventions for the home are legion. The now ubiquitous vacuum cleaner was once derided as absurd; likewise electric light, which was seen as a hopelessly futuristic alternative to gas and oil lighting. Other 'crazy' inventions are more recent – the second half of the twentieth century produced chopsticks joined to a fork (so you could have the fun of using chopsticks without the bother!), an inflatable hearthrug and sunglasses that attached to the wearer's head using two magnetic discs permanently stuck to each side of the head.

But if that's the maddest, most apparently useless device, what of the pat-on-the-back machine which won approval from counsellors and psychiatrists worldwide? Filed for patent in 1936, it involves an elaborate contraption strapped to either the left or right shoulder. Suspended from part of the contraption so that it can be seen dangling in front of the eyes and in close reach of the hand is a thin chain or piece of strong string. When pulled, a series of levers and hinges is operated, engaging another rod running down the wearer's back, which is attached to an artificial hand that pats the wearer on the back. However often you pull the string, the back patter never gets fed up or tired. Pull hard and you get a firm pat; more gently and the back patter gives you what almost amounts to a reassuring hug. The device, though never put into commercial production, really did meet with approval among psychiatrists who thought it would promote well-being and self-esteem among the depressed.

SOMETHING'S COOKING

ENGLAND, 1945

The realisation that visible light is just a narrow part of a wide band of electromagnetic waves/particles was one of the most remarkable discoveries in the whole of scientific history. At one end radio waves are really long; then come microwaves; then infrared; then the visible spectrum (the part of the electromagnetic spectrum that we humans can see); then ultraviolet (ultraviolet waves are too short for us to see, but visible to bees and other insects); then X-rays, which are so short and high energy they go through you; finally we reach gamma rays – extremely short, very high energy and extremely dangerous as they disrupt cells and cause cancer.

During the Second World War scientists used their knowledge of the electromagnetic spectrum to improve Britain's radar system by adding what became known as a magnetron. The magnetron produced microwaves, which made radar systems much more effective, but one scientist noticed – quite by chance – that microwaves were also very good at something else.

The curiously named Percy Le Baron Spencer realised that microwaves could in fact heat food far more rapidly than conventional food-heating systems like ovens – one of the more unexpected spin-offs from wartime research.

Spencer's company Raytheon knew they were on to a good thing, however, and by 1954 they had built the 1161 Radarange which was a big, very expensive but also highly effective microwave oven. That first microwave oven was seen as a bit of a freak by the general public, something that could never catch on, but increasing improvements in design meant the price gradually fell and microwave ovens could be sold in a more manageable size; within thirty years the microwave oven, an accidental by-product of something entirely different, was almost as common a household item as the kettle.

For the technically minded, light waves within the visible spectrum have a wavelength of 400–700 millimicrons; microwaves have a wavelength of 0.1 millimetre.

STICKY BEGINNING

SWITZERLAND, 1948

For thousands of years after early man left the warmth (and occasional famine) of Africa to populate the rest of the globe, clothing was a problem. As mankind moved north to inhabit the colder regions of the earth, clothing became essential. We don't know how animal skins were fastened round the body or even if they were fastened at all in very early prehistory, but by the Greek and Roman eras brooches of various kinds and leather thongs were certainly being used on increasingly sophisticated clothing. But that was it more or less until the development of the button, the zip and then – perhaps the most bizarre in its origins – the Velcro strip.

Like countless generations before him, George de Mestral enjoyed going for a walk in the countryside. De Mestral was also an extremely observant scientist and he noticed that every time he returned from a certain favourite route near his Swiss home, his clothes were covered in burrs that were extremely difficult to remove.

Intrigued, de Mestral examined the burrs and discovered something quite amazing: they were covered in a dense layer of stiff hairs that ended in a distinct hook. It is said that de Mestral immediately conceived the idea for a completely new kind of fastener, but if the idea came quickly the practical application took a little longer.

In fact, it took eight years of hard work to produce two strips of nylon fabric that would do what nature seemed so effortlessly to do with burrs. De Mestral tried dozens of different materials but it was only a particular form of nylon that worked. The successful design, which was finally achieved in 1948, consisted of a strip of nylon with densely packed hairs ending in loops and an opposing strip of nylon covered in densely packed hairs ending in hooks.

What is really extraordinary about the material and the design is that the strips can be attached and torn apart almost indefinitely without loss of effectiveness. De Mestral called his new invention Velcro and although many people were sceptical and thought it would never catch on in a world already richly endowed with buttons, zips and hooks and eyes they were wrong: Velcro was quick, easy and cheap to make and is now common right across the world.

ALL GUMMED UP

AMERICA, 1949

The American passion for chewing gum baffled Europeans when they first came upon it during the Second World War, but now the habit is almost universal – from Helsinki to Beijing youngsters are addicted to the stuff to the bafflement of their elders and, occasionally, to the irritation of the civic authorities who have to find ways to scrape used gum off roads, footpaths, shop floors and park benches. For if there is one thing that chewing gum is very good at, it's sticking to things.

American inventors have long been aware of the problems of disposing of chewing gum and at least one – Reva Harris Keston – came up with a noteworthy device for dealing with it. In the years after the Second World War, Ms Keston invented what she called the used gum container. It sounds disgusting but is actually far worse even than it sounds because the point of her invention is that the container is designed to hold discarded gum but also gum that you may have chewed for a while but want to keep for later on!

The container – which was patented with all due seriousness – is actually a flat piece of cardboard with a flexible pointed sliver of metal sticking up from the middle of it. The idea was that you bought them in flat packs of a dozen or more – one for each piece of gum you intended to eat. Each piece of cardboard was folded along pre-marked dotted lines so that when the time came you could easily turn it into a little box, the metal spike, which was folded flat against the cardboard when you bought your container, was bent upwards when the thing was in use.

So if you were halfway through your gum and wanted to stop chewing while you had dinner, you bent the metal spike until it was sticking up at right angles to the cardboard and then impaled your gum on the spike until you felt like going back to it. If you'd finally finished with your gum, you left the spike down, placed the gum on the cardboard tray, folded along the dotted lines to complete the sides and top of the box and then threw the whole thing in the bin!

Ms Keston never seems to have considered the possibility that people would find the idea of keeping used gum for re-chewing later rather disgusting, nor that it was unlikely that people would want to buy a dispenser for gum they traditionally chewed and then simply threw away.

This was not an invention that would change the world.

ULTIMATE REALITY

ENGLAND, 1949

It's easy to laugh at the idea of pure physicists who study the distant universe and tell us, among other things, that our solar system is moving away from all the other solar systems at a constant 200 miles per second or that even the most solid pieces of matter are actually mostly empty space. It all seems to have nothing to do with our daily lives and even if it's true, so what?

But that is to miss the point because many high-energy physics inventions have actually had spin-offs that affect all our lives. Huge particle accelerators might seem meaningless to most of us but what they do is actually an extension of what a TV does. A TV is, if you like, an extraordinary invention that fires streams of particles at a cathode ray (the screen) and each one bursts into a pinprick of phosphor light; put enough pinpricks together and you have a television picture.

Moving from small domestic particle accelerators like TVs to the giant accelerators used by physicists takes us into a world as strange as anything in science fiction. The first particle accelerator was the brainchild of Ernest Walton and Douglas Cockcroft, for which the two men shared the 1951 Nobel Prize for Physics.

Half a century later, particle accelerators can be vast – more than a dozen miles long in some cases. They run through deep underground tunnels and are hugely complex and expensive to build. They are lined with pure copper and use vast amounts of energy to produce what no one can actually ever see.

It is not easy for non-scientific minds to understand exactly what a particle accelerator does and why but the basic idea is that it uses high energy to speed up particles smaller than an atom and then aim those particles at a special target.

The really big high energy accelerators can drive particles along at speeds close to the speed of light (and one of the basic laws of Einsteinian physics is that nothing can go faster than the speed of light). When they reach these high speeds and hit specially made targets the particles are smashed into smaller particles.

So powerful are these accelerators that we are now well past the days when atoms and their nuclei were considered the smallest pieces of matter in the universe.

Particles are accelerated by electromagnetic waves and the collisions that occur when they are smashed create radiation but also provide information about the most elementary structure of matter. Subatomic particles so far detected include fermions (these make up known matter and antimatter), leptons (including electrons and neutrinos), quarks, and antiquarks and antileptons (so-called antimatter).

At the time of writing, quarks are still officially the smallest particles known, but bizarrely – and this is what gives the particle accelerator its strange reputation – they have never been seen. Along with all the other subatomic particles they are detected using liquid gas detectors and other devices which record a trace as the particle moves through them. So what we see is the path taken by the particle rather than the particle itself – and even to see this we need millions of pounds' worth of equipment built into an underground tunnel system that may run for twenty miles!

The more one looks into the particle accelerator the more remarkable it becomes – the lifetime of many of the particles the accelerator was invented to detect, for example, amounts to less than a billionth of a second, but the discovery of the ultimate particle will almost certainly depend on building even bigger and more powerful accelerators.

SILLY PUTTY

AMERICA, 1949

The disruption in world trade caused by the Second World War meant that rubber was very difficult to come by. For the Americans, once they'd entered the war, this was a serious problem because they needed rubber for the war effort – no rubber meant no boots for the soldiers, no tyres for aeroplanes, trucks and cars.

Ever inventive, the government mobilised the American scientific world to see if they could come up with a substitute for rubber, and it was quickly discovered that one of the most promising areas of research was – bizarrely – sand. Sand was a fruitful area for researchers because a major component in sand is silicon and silicon seemed to have many of the properties that make rubber so useful. The difficulty was extracting the silicon and then finding a way to harness its useful properties while eliminating its less than helpful qualities.

James Wright, an employee at General Electric, decided to see what would happen if he added boric acid to silicon oil. The result was extraordinary. Wright had produced a strange gooey mass that held together despite looking as if it couldn't, and which bounced. It was completely useless from any practical point of view and certainly couldn't be used to make tyres or boots, but that didn't stop General Electric and a host of other companies worldwide trying to find something to do with it. After years of fruitless efforts its interesting properties remained undiminished but it was given up as a dead loss from every practical angle.

Then, in 1949, a toy-maker called Peter Hodgson decided that everyone had been thinking of silicon goo from entirely the wrong angle – if it had no serious use perhaps it could have a frivolous use. Thus Silly Putty was born.

Not only did the goo bounce, it was also very good at picking things up and could be made in dozens of colours. Hodgson was so convinced that goo would be successful as a novelty toy that he borrowed enough money to start production and within a few years had created one of the most popular toys of all time – a toy that became a huge fad among adults and children.

Silly Putty goo was sold in plastic eggs and it has continued to sell since it appeared on the market in 1949.

Ironically, after being rejected as an invention with no practical application the advent of silly putty proved that it did have some vaguely practical uses after all – for example, astronauts used it to hold tools and pens in place in their weightless environment; it was marketed as a highly efficient tool for removing dust and hairs from fabrics; tennis players and other athletes used it to squeeze continually in their hands in order to strengthen their wrists; it was even used as a mould – at least one American zoo used it to take imprints of gorilla feet!

PIE THROWING

AMERICA, 1950

It's hard to believe that anything as mundane as a cake tin could be transformed into one of the most popular of all sports toys. The story begins with a baker called William Russel Frisbee.

Mr Frisbee made pies and sold them in thin metal pie dishes that could be reused by the pie-buyer if he or she liked, but were cheap enough to be thrown away. Some time in the late 1870s Mr Frisbee had a brainwave – he thought that the company's fortunes could do with a boost so he decided that it would be a good idea to put the company name on the bottom of each and every metal cake tin. It would be cheap to do and provide a good way to advertise the company at low cost.

By 1950 Mr Frisbee's splendid pies were still being made complete with their lettered tins and in that year Walter Frederick Morrison, who was known as a flying saucer enthusiast, decided that keeping pies in place was a poor use for something so aerodynamically interesting as a Frisbee pie dish. He'd seen students at Princeton University throwing the tins across the quad and been inspired by their ability to fly.

He made a few adjustments to the shape of the tin to improve its speed and lift and the modern Frisbee was born. He set up the Wham-O company to make, market and sell Frisbees and in the succeeding half century and more the Frisbee has taken the world by storm, with billions of the shallow discs – now made in plastic – being sold all over the world.

WASHING WHEELS

AMERICA, 1952

The American love affair with the car seemingly knows no bounds. After Henry Ford came up with the Model T – the first car to be mass-produced – cars gradually became something every American simply had to have. But once you have a car the next thing is to gradually add as many refinements and gadgets as possible. Over the years American cars increased dramatically in size and complexity and long before other countries in Europe followed suit, American cars had air conditioning as standard, electric windows, cruise control, sunroofs and much more. American companies are nothing if not competitive, and the desire to make a particular car more appealing to the customer led to endless refinements and additional gizmos, some of which were – to put it as politely as possible – eccentric.

Take the strap-on travel washing machine, for example. This carefully thought-out device was designed to make camper travel as good as staying at home. Americans love their home comforts and taking a fridge and a washing machine on camping expeditions doesn't seem the least bit odd on the other side of the pond. The travel washing machine was invented and patented in the early 1950s when the passion for driving was probably at its height in America. Fuel was still cheap, cars and camper vans were huge and powerful, and getting away from it all to some remote backwoods location was all the rage – just so long as it didn't mean leaving any labour-saving devices at home.

The patent drawings for the travel washing machine reveal a meticulously – even brilliantly-engineered drum within a drum. The whole thing strapped to the rear wheel of the car or camper van so that the net result was just an unusually fat wheel. Once the machine was fixed into position, its glass front door was opened and the clothes and washing powder inserted. An opening at the top of the drum was then used to fill the machine with water. All you had to do then was drive off and, as the wheel turned, so too the washing-machine drum turned and – hey presto – half an hour later your clothes were washed and you'd travelled fifty miles.

Unfortunately, that's where the complications come into the picture. After half an hour's driving you had to stop to drain off the washing water and then fill the drum again so the clothes could be rinsed. That meant another half-hour's driving. Worse still, the car or camper van had to be jacked up each time you wanted to attach or remove the washing machine. The final – and ultimately insurmountable – problem was that during trial runs it was discovered that travelling at more than 25 miles an hour virtually destroyed your clothes. The buttons and zips were mangled and the G-force inside the drum was so powerful at, say, fifty or sixty miles an hour that clothes were stretched or compressed until they could hardly be rescued. At eighty miles an hour the clothes started to become torn.

However, the idea of having your washing machine while camping was too compelling to be completely dropped and eventually machines were introduced that ran off the car battery or used the power generated by the engine. By this time travel fridges, travel lavatories, travel TVs and travel satellite communications were becoming almost standard.

IN THE DEEP FREEZE

AMERICA, 1962

Some inventions attract people despite the strongest evidence that they are pointless and a waste of money. Perhaps the best example of this sort of thing is the invention of cryonics by C W Ettinger in 1962, the theory being that if you freeze people when they die some future technology may be sophisticated enough to bring them back to life with the personality, the memories and the consciousness they had in life.

Cryogenics certainly works in the sense that at ultra low temperatures physical decay can be halted. Deeply frozen bodies can last for thousands of years so long as they don't get accidentally warmed up.

American cryogenic centres offer deep-freeze facilities with all sorts of power failure back-ups guaranteed to ensure so far as humanly possible that neither flood nor earthquake (nothing short of a full-scale nuclear attack, in other words) will interrupt the supply of electricity to clients who have paid huge amounts to be frozen in the hope of eventual resurrection.

The technique, which developed gradually after the Second World War, involves treating the recently dead with a chemical that prevents ice damage to tissue and then gradually reducing the temperature of the cadaver using liquid nitrogen. In other words, cryonics doesn't involve a simple freezing process – the sort of process that takes place when you put food in the deep freeze – instead it uses much lower temperatures to halt any process of decay.

Some scientists – and even some serious ones – believe that eventually technology will advance to the point where any amount of cell damage will be reparable – which would mean that a thawed corpse, despite the freezing process and the illness or injury that caused the original death, could be defrosted, repaired and brought back to life. The belief that this will become possible is usually based on the fact that many commonplace scientific inventions were once considered utterly and completely impossible. Eighteenth-century scientists, for example, insisted that humans would never ever be able to travel at more than thirty miles an hour; flight was, until the late nineteenth century, beyond any imagining; and belief that people would one day land on the moon would, in earlier ages, have resulted quickly in admission to a lunatic asylum.

The cryonics enthusiasts insist that what they are doing falls into the same sort of category and there is no doubt that developments in genetic technology suggest that we will sooner or later be able to adjust the human body on a molecular level. However, the difficulties are formidable, as one eminent scientist explained:

> First, there are fractures in the frozen tissues caused by thermal strain – if we warmed our hero up, he'd fall into pieces as though sliced by many incredibly sharp knives. Second, suspension is only used as a last resort: the patient is at least terminal and current social and legal customs require that the patient be legally dead before suspension can even begin. While the terminally ill patient who has refused heroic measures can be declared legally dead when he could in fact be revived (even by today's technology), we're not always so lucky. Often, there has been some period of ischemia (loss of blood flow), and the tissue is nowhere near the pink of health. The powerhouses of the cells, the mitochondria, have likely suffered significant damage. 'Floculent densities' (seen in transmission electron microscopy) likely mean that the internal membranes of the mitochondria are severely damaged, the mitochondria themselves are

> probably swollen, and cellular energy levels have probably dropped well below the point where the cell could function even if all its biochemical and metabolic pathways were intact. The high levels of cryoprotectants used in the suspension (to prevent ice damage) have likely poisoned at least some and possibly many critical enzyme systems. If the cryoprotectants didn't penetrate uniformly (as seems likely for a few special regions, such as the axonal regions of myelinated nerve cells: the myelin sheath probably slows the penetration of the cryoprotectant) then small regions suffering from more severe ice damage will be present.

Despite all this, cryogenics attracts both scientists and those who are willing to be frozen, which is understandable given that any invention – however outrageous – that promises even the slimmest chance of eternal life is bound to have its fans.

ON THE SLIDE

SOUTH AFRICA, 1963

It's perhaps ironic that a man who grew up in South Africa, a country where it almost never snows, should come up with an astonishingly detailed (and loopy) idea for a wheel made of ice.

Dragan Rudolf Petrik may have had a holiday in some snowbound country and, excited by this strange new phenomenon of ice and snow, thought he really had to try to do something useful with it. The sliding properties of ice being well known, it perhaps seemed like a Eureka moment when he came up with his design for ice block wheels.

The best description of an invention that would almost certainly be otherwise indescribable comes from the patent application itself. Here Mr Petrik explains his core ideas:

> The present invention relates to a means of transport comprising the application of ice blocks in any form, shape and size as skids for any kinds of vehicles, such as cars or trains which are driven by propellers, jet engine electrical or other power, which ice blocks slide on smooth metal tracks or other prepared surfaces, which can be heated, which sliding blocks are always for a constant distance extended out of the guide ducts, which distance is automatically regulated by a conventional electronic eye device, which is coupled to an electro-motor which rotates spirally threaded bearing columns,

which rotation pushes coupled press pads and so also the ice blocks in guide ducts down against the track for the required sliding promoted by melting of the extended part of the ice block, which part of the ice block is cooled with cold air from an air blowing device comprising a freezing machine is situated in the store room for the ice blocks.

The sliding ice block when exhausted being replaced by a reserve ice block which is pushed down the track through a reserve guide duct in the same manner as the previous ice block, the vehicle so travelling being supported by the new ice blocks, exhausted ice block guide ducts being refilled with reserve ice blocks, which are stored in a work room which is cooled by a freezing machine, and to which room replenishing ice blocks are brought through an entrance situated in the ceiling and wagon mantle construction wherein a so equipped vehicle travelling on curved road sections is protected against centrifugal derailment by means of a vertical flange wall on the tracks and a safety fence against which are supported bearing pads on the vehicle.

So that's clear then.

DESIGNED TO BREAK

AMERICA, 1963

Anyone who enjoys either playing sport or watching it on television will have seen or experienced that terrible moment when, after a day of disasters, the mind goes into overdrive and the sportsman or woman turns into a raging, crazed animal bent on destruction. Certain well-known tennis players are prone to this kind of tantrum and the result can be screams, kicking, frothing at the mouth and, at worst, a broken racquet.

Other sports are not immune to these complete losses of control. Horrified at the damage to the sport's reputation, one golf enthusiast and amateur inventor came up with a sure-fire answer to what he saw as an epidemic of expensive golf-club smashing: he created an authentic-looking selection of clubs – putters, drivers and woods – that looked exactly like the real thing but were actually cheap and designed specially to break. In fact, that's all they were designed to do; if you tried to take a swing at a ball with one it would snap in your hands. Useless for golf proper, they were perfect for golf tantrums. The designer even came up with a golf bag with a slightly separate compartment in which the sportsman was supposed to carry two or three temper tantrum clubs. The separate compartment reduced the risk of the irate golfer whipping out a genuine club and smashing it to bits.

The Breakable Simulated Golf Club – to give it the official name that appears on its official patent – was remarkably realistic-looking and its inventor was sure it would make him a fortune. Sadly the golfing world did not take kindly to the idea that it was riddled with players inclined at the drop of a hat to go berserk. Madmen and lunatics they might be, but they were none too keen to advertise the fact to the world. There was also a suspicion among some golfers that the only real point of losing your temper after a bad shot was to have the pleasure of smashing a real and very expensive golf club. Smashing a cheap, flimsy little club held no real pleasure for these giants of the fairway.

FALLOUT TENT

AMERICA, 1963

Einstein's famous equation $E=mc^2$ is generally taken to mean that matter and energy are the same thing from a scientific point of view. Anyone who doubts this need only remember the bombs that dropped on Nagasaki and Hiroshima – each depended for its awesome destructive power on that simple equation. A couple of dozen pounds of specially enriched uranium in each bomb was triggered in such a way that all the neutrons within it – and that means countless trillions of neutrons – smashed into each other in an unstoppable chain reaction that took the big lump of metal out of existence, transferring it to pure energy. There was no explosion in the conventional sense; no TNT or gelignite could produce anything remotely near the power of the chain reaction in an unstable lump of metal.

Even for the scientifically literate, the difficulties of understanding the power of atomic and later nuclear energy, the mechanism by which it works and the science on which it is based were formidable. The problem was that nuclear and atomic explosions had almost nothing in common with conventional explosions however big. Conventional explosions use only a tiny fraction of a material's rest-mass energy (to use the scientific jargon) by rearranging, however violently, the atoms in the materials. A nuclear explosion uses a much greater part of that energy by releasing the energy latent in the atomic structure of matter itself.

And all this may explain why an otherwise intelligent American inventor could come up with a portable nuclear fallout shelter and market it as a serious solution to the threat of possible nuclear war.

Invented in the 1960s, when the Cold War was at its height and Americans were convinced that at any moment they would be subject to a pre-emptive nuclear strike from the USSR, the fallout shelter consists of an ordinary tent with extra holes for ventilation, a built-in water tank and a hose pipe.

The instruction manual for the shelter tells the buyer to dig a trench in the garden and then pitch the shelter above the trench. The hose pipe can be fed down into the trench where the family sit while the war goes on above them. So long as they take plenty of tinned food they should be able to ride out the storm. One of the nicest things about the shelter is that the owner is instructed to flap the sides of the tent to increase ventilation.

With the truth about atomic and nuclear weapons known only to a relatively few, how could buyers of the shelter have known that nothing can withstand the power of nuclear detonation and even if you happened to be hundred of miles from the site of the nearest explosion, radiation sickness would quickly kill you and all other life forms. That a flimsy tent pitched over a hole in the ground should be thought capable of withstanding a nuclear war is in a sense a tribute to an innocence about such matters that is now long gone.

BOTTOMS UP

GERMANY, 1965

The list of patents attributable to German scientific inventors does not do a lot to enhance that nation's reputation. The stereotypical German obsession with cleanliness is confirmed by two hilariously mad inventions to do with bodily functions. In the 1960s a German scientist sent a long, detailed patent for a bottom-spreading lavatory seat.

It is hard to believe that anyone would produce detailed engineering drawings to create such a thing, but this complex seat was designed to move outwards a carefully measured amount when the lavatory user sat down; the amount of movement was proportionate to the weight of the sitter. The aim was to make defecation cleaner, easier and more comfortable. The loo seat, designed using high-quality precision ball bearings and wood and metal machined to fine tolerances, was extremely expensive – a fact that led to the claim that only a German would be crazy enough to pay for such a thing.

Not content with showing themselves up with the bottom spreader, the Germans then revealed their next anally fixated invention: a flatulence collector. This consisted of a tube (for insertion into the anus) and an airtight bag. The circumstances under which this might be used – it is recommended for use in the home and the office – do not bear thinking about.

Last but by no means least there is the patent dog excrement gatherer, again a German invention. The idea is that rather than wait for your pet pooch to soil the local park or garden you insert the Go Matic device into your dog's bottom before you take it for walkies. The tube, decorously surrounded with a protective funnel so you don't have to see what you're doing, is inserted in the dog's rectum and a small foot pump is then attached to the free end of the tube. The user operates the pump until the dog's bowels have been emptied down the tube into a bag.

What the dog thought of this undignified device is not recorded.

BUNNY WITH A STING IN ITS TAIL

AMERICA, 1967

All over the world parents dread taking their children to the doctor. With the exception of doctors in the British private health system, the problem is not that doctors are themselves ferocious or unpleasant in any way. The difficulty is that for children doctors are associated with just one thing – injections – and for most if not all children the prospect of having an injection is enough to put them off doctors and hospitals for life.

Ever quick to see an opportunity, American inventors and medical scientists thought there had to be a way round the problem. Getting the mother to talk to the child and popping the needle in when the child is least expecting it was tried but the results if anything were worse than telling the child he or she was about to be injected.

You might be able to inject a child off guard once but the next visit to the doctor would be likely to be extremely traumatic as a result, with the child (assuming it could be got to the surgery) constantly on the lookout for sneaky medical behaviour. At last one inventor was convinced he had the solution and in 1967 the Bunny Syringe was invented. Odd-looking in the extreme, the Bunny Syringe is exactly that: a syringe cunningly concealed in a plastic bunny rabbit. Of course the bunny is designed to work in hospitals and clinics so it only looks like an ordinary toy bunny. In fact it was designed with a built-in syringe that used the bunny's tail as a plunger and the body as a reservoir for the medicine being administered.

The only terrifying giveaway was the bunny nose extension – actually the needle. No one knows if the Bunny Syringe worked, but the fact that it never caught on suggests that it wasn't such a great idea after all – in fact, it would probably only give children a horror of bunny rabbits!

ALARM BALLS

AMERICA, 1968

Every film that's ever been made about golf includes a scene where the golfer loses his or her ball. For the professional golfer of course it's a relatively minor risk but for the amateur it is one of the most annoying and undignified aspects of the game. That long dreamed of perfect drive somehow doesn't work as planned and the ball sails off into the trees where the golfer spends an age fruitlessly searching for it only to have to start all over again while his or her colleagues wait impatiently up ahead.

There had to be a solution – and there was: the fully patented alarm golf ball.

Because of the forces involved – the enormous compression and acceleration caused by being hit by the club – the alarmed golf ball was tricky and expensive to design and make, but early prototypes at last seemed to work well. The ball would be primed at the first hole and then hit. If all went well it would act just like an ordinary golf ball. If at some stage it was miss-hit and entered the rough and after a reasonable search could not be found, the golfer activated a switch on a small electronic device which caused the golf ball to emit a regular alarm signal. The golfer could then march straight over to the ball and resume playing. So why didn't the smart ball catch on? Well, like so many fine (if rather odd) inventions its cost was out of all proportion to any benefit it might offer. It is annoying, golfers agreed, to lose a ball in the rough, but not so annoying that they would happily pay five times the cost of an ordinary ball just to sort out the problem.

An even madder idea than the relatively sane alarmed golf ball was patented in the same year. This was aimed specifically and unashamedly at keen golfers 'with very little talent or experience'. The people who marketed the homing golf ball insisted that it gave harmless pleasure by working in tandem with an electronic device that had to be placed in the hole before the player teed off.

On a hole that should have four strokes to complete, the homing golf ball would act perfectly naturally until it reached the green. If it landed on the green within twenty feet of the hole and was travelling roughly in the direction of the hole, the golfer could activate ball and hole so that the ball was drawn irresistibly to the hole.

It might have been cheating, but golfers were encouraged to see the device as a confidence booster that should only be used on genuine practice rounds and not in competition.

BLANKET SOLUTION

ENGLAND, 1969

After the horrendous destruction of the Twin Towers in New York, people are understandably nervous about living or working in skyscrapers. It isn't just the risk of terrorist attack that makes people worry – the other great danger is fire. Although fire emergency procedures are in place for most large buildings, in practice it is very difficult to evacuate thousands of people from a really high tower block.

The great British inventor (none of whose patents was ever put into commercial production) Arthur Paul Pedrick turned his mind to this problem and came up with what on paper looks like a giant collar around the building somewhere near the top. The collar is actually a huge piece of rolled-up material running right around the building. Made from a fireproof material, the collar was designed so that it could be attached to the electronic sensors that make the sprinklers work. Should a fire start, the sprinklers would activate in the normal way and then the fireproof collar would unravel, running down the sides of the building until the tower block was completely enveloped as if in a giant sock. The sock would immediately stifle the fire and protect the building and its inhabitants.

To the objection that the people in the building would quickly be deprived of oxygen, Arthur Paul Pedrick would have replied that he'd thought of that. His blanket was designed to be fitted with small holes in certain places and fire officers inside the stricken building would direct the occupants to these apertures. There would be enough oxygen for breathing but not enough to encourage the fire.

In theory, of course, it's a great idea but the practicalities of getting the blanket in position, ensuring it unrolls quickly enough and isn't damaged by long periods during which it is never used proved insurmountable.

DOG EARED

AMERICA, 1970

The science of dog breeding has thrown up some pretty peculiar-looking creatures – from hairless hounds to pocket-sized pooches, from giant slobbering scent dogs to lumbering St Bernards. Some breeds are now so odd that they can't copulate without assistance (dachshunds) or breathe properly (pugs), but having gone to all the bother of creating these weird breeds people then started to come up with crackpot ideas to help dogs overcome their inbred problems.

One American inventor designed a lightweight pair of wheels to help long, thin dogs that are prone to severe back problems; as the dog gets older and loses the use of its back legs you strap on the wheels and the dog gets a new lease of life. Indeed, the world of dogs is filled with barmy inventions – there are dog trailers to tow behind bicycles, dog Walkmans for mutts who like music, dog exercise wheels (like a giant hamster wheel), do-it-yourself dog hair dye and perm kits, and even a special dog hat that is fitted with a small umbrella.

Perhaps the silliest dog device, however, is the dog ear protector. This was designed by an American inventor who'd noticed that his King Charles spaniel always managed to drag its long, floppy ears through its food, resulting in a mess on the floor and an animal that had to be cleaned up. The solution was a hat that strapped round the dog's head and was fitted with two tubes. With the hat strapped firmly in place the two tubes stood out at right angles either side of the dog's head and the long ears were pushed one through each tube. The inventor doesn't seem to have been bothered about the fact that this elaborate bit of headgear would have to be fitted before each and every meal and then removed afterwards. The cleaning up process would almost certainly be far quicker!

ON YOUR KNEES!

AMERICA, 1972

Fifty years ago all skis were made from laminated wood. They were heavy and they needed regular attention and varnishing to avoid splitting and generally falling apart at the seams. Then along came fibreglass and modern lightweight skis were born. That revolution was followed by even lighter, stronger materials such as graphite and boron, which now produce incredibly light and flexible skis that need almost no maintenance.

The basic design of the ski remained the same until some bright spark thought of the mono ski; if you're a really good skier the mono ski simply confirms your ability to ski with your feet clamped tight together in a style that experienced skiers often try to adopt with two skis anyway. But then came the snowboard and skiing was transformed into a sport split into old school – conventional skis – and new school – snowboards. Snowboarding became the thing to do if you were young – a badge of identity to distinguish young, cool skiers from their staid parents.

By now the inventors had turned their attentions to skis and skiing in a big way, which meant a plethora of weird ski inventions. First there were tiny skis (about eighteen inches long), which were said to be easier to control for beginners than the traditional long ski. Then came what has to be the nuttiest ski invention ever: shin skis.

Invented in the early 1970s these bizarre skis were strapped one to each leg, but not in the usual way. The shin ski started just above the knee and ran down the shin against which it was tightly moulded and held in place by straps. At the instep it doubled back on itself and ran back up the leg a few inches forward of the leg itself and stopping with a traditional ski tip curve halfway up each thigh. Each set of shin skis came with two tiny hand skis fitted with hand grips.

The idea was that you crouched down and rested on your shins, with your hands (each holding a tiny hand ski) out in front of you and out in front of the shin skis. The thing about the shin ski is that it worked. You could happily hurtle down the piste using your hands to help steer and leaning from side to side to make your turns. When you came to a stop you simply stood up and the skis were no longer a problem – you didn't have to take them off and carry them in the usual way because they were parallel to your legs and your feet were now on the ground (remember the skis ran round your instep not under the soles of your feet).

The main problem with shin skis and the probable reason for their demise was said to be that they made you look like an idiot – as if you'd strapped a child's toboggan to your front. So, unlike the snowboard, the shin ski never caught on.

THE VANITY OF HUMAN WISHES

AMERICA, 1973

The vanity of human beings knows no bounds. From injecting rat poison into our wrinkles to sewing hair to our heads (via plastic bags in our breasts, gelatin in our lips and implants in our buttocks) we are prepared to do almost anything to regain something of the beauty we once had or to which we feel we are entitled.

Before cancer scares made sunbathing not so much a fad as a serious health risk the tanned look was absolutely vital for those who really wanted to be noticed – in America, research suggested it could even add greatly to your chances of promotion in your job.

The great difficulty was how to get and maintain an even tan. In the colder states of east coast America, tanning salons sprang up almost everywhere and eager customers were prepared to lie regularly under what look like giant toasters to achieve the perfect orange look.

On the warmer West Coast the tan was everything and no one would dare to appear on the beach without a toned body and a good, even tan. Tanning was easier here too because the sun shone most of the time, but then lying in the sun is no guarantee of an even tan: what was the good of having a dark back and a pale front, or vice versa? Such a tan looked like incompetence or a lamentable lack of self-respect.

All of which may explain why numerous exotic contraptions were invented to help the tanners – aluminium reflectors were made and strapped to the chin (to reflect sun upwards as well as downwards on to the face), little nose protectors were created to prevent that prominent part of one's anatomy getting burned, and then in 1973 came the silliest of all tanning inventions: the toe-ring.

Invented by Russell Grathouse in 1973, the toe ring was designed to help ensure an even suntan on the feet. Mr Grathouse had identified a serious problem in conventional suntanning techniques: if you lie on your back, your feet tend to splay out sideways which means that the inside edge of each foot tends to get more sun than the outside edge. This wholly unacceptable state of affairs was rectified by Mr Grathouse's rings. These consisted of a pair of joined plastic rings, one to be placed over the top of each big toe, thereby preventing the feet moving apart.

As a final delicate touch Mr Crathouse provided a much smaller ring-sized hole between the two toe holes. This was designed to hold a flower while the rings were being worn.

ATOMIC CATFLAP

ENGLAND, 1974

The great Sussex inventor – and former employee of the Patent Office – Arthur Paul Pedrick spent his retirement in the 1970s thinking up wonderful (if slightly impractical) inventions: everything, in fact, from a system to irrigate Africa by firing rocket-propelled snowballs from Antarctica to a catflap designed to communicate with an atomic bomb orbiting above the earth.

The catflap was developed to keep out any cat other than Mr Pedrick's own ginger moggy. It worked by using a colour sensor that was sensitive only to ginger – black cats and tortoiseshells wouldn't be recognised and the catflap would remain firmly closed.

The best part of Mr Pedrick's patent application comes where he records his detailed conversations with Ginger (the cat in question), who clearly helped enormously with the detailed process of catflap invention.

Why the catflap should be attached by radio transmitter to an atomic bomb in space remains a mystery and we have no idea if the catflap was ever actually installed as Mrs Pedrick soon put a stop to her husband's compulsive inventing – he had filed more than fifty applications and she decided enough was enough. The silence from Sussex must have been deafening!

MUD SHOES

AMERICA, 1974

Scientists come in two basic types – the ones who investigate highly esoteric fields of enquiry which may or may not eventually produce something of practical value and scientists who concentrate entirely on solving practical problems that have afflicted mankind for centuries. Of course the difficulty comes when an obsessive or eccentric scientist gets a bee in his or her bonnet about something and ends up spending years trying to perfect a device or contraption that is designed – even when perfected – to solve only a minor problem. This is the point at which the practical scientist meets the pure scientist – a device designed to eliminate dust may be useful but is it worth creating something vastly complex and expensive to do it? The practical scientist who is beginning to turn into a pure scientist will reply that he or she doesn't care – the challenge is to make the thing work however much effort, time and cost is involved.

And of course scientists and inventors tend to some degree to believe in their inventions if they have spent a long time on them – which may explain why a system of shoe mudflaps was patented at great cost in the early 1970s by an American inventor who surely must have realised that practical issues are not uppermost in the minds of fashion-conscious shoe buyers!

The shoe mudflap did, however, address a real problem. How many women have been irritated to discover that the process of walking across a wet street has thrown a great deal of dirty water up the backs of their legs? Not only do one's stockings get in a mess but so do one's shoes.

The answer, thought our intrepid scientist, was to come up with a set of mudflaps for shoes that would fit all kinds of women's shoes, from the fearsome stiletto to the cumbersome platform shoe.

The mudflap was a cleverly shaped and fitted plastic device, available in various colours or see-through, that clipped to the shoe just at the point where it reaches the ankle. Apart from the expense and the fact that the mudflaps made every wearer look as if she was sporting wings on her ankles, the biggest problem with the shoe mudflap was that it was very difficult to make it work on the vast range of shoe types available – and that meant coming up with a range of designs almost as extensive as the shoes themselves. And so it was that the shoe mudflap slipped quietly into history. A practical idea that was, in the event, completely impractical.

SHOCK TREATMENT

RUSSIA, 1975

You can never be sure with strange inventions – they start off seeming mad and end up being central to people's lives. The microwave oven, for example, or the horseless carriage, 'That will never catch on, not in a million years!' said a high court judge in 1902. Famous last words indeed.

On the other hand, many inventions are based on sound principles but somehow they still just don't seem to work or even in some cases to make any sense. Everyone has heard about the Pavlovian principle which is based on the idea that responses can be conditioned. The classic experiment involved ringing a bell and then feeding a dog. After a while the dog begins to salivate every time the bell rings whether or not he then gets fed. Of course if the bell continues to ring and the dog never gets fed the response starts to fade.

A Russian inventor during the Soviet era decided he would create a device to encourage learning and memory retention that was based loosely on this Pavlovian idea. The inventor is reported to have presented his idea to a senior local government official who passed it on to someone in the Politburo, the Soviet Union's supreme governing body in those bad old predemocratic days.

The inventor immediately received a letter congratulating him and offering him full funding for the project. A response of that kind was almost unheard of in the USSR, however pressing the need for a quick decision. So what was the invention that captured the minds of those hardened communists? It was called the electrical stimulator. The idea was that children and adults being taught science, history, maths, chess or whatever would be wired up to the device during their lessons. If they were supposed to have learned something the device was activated during a test. If, in response to a question, the student gave the wrong answer the tutor would press a button on a console and the device attached to the student's arm would immediately transmit a mild electric shock. If the student persisted in getting the answers wrong the strength of each subsequent electric shock would be increased. The inventor insisted that the device was so effective that seriously high voltages would never have to be used, but for the squeamish or over-sentimental (as the inventor reportedly put it) he offered to include a fail-safe mechanism that would establish a maximum voltage.

The inspiration for the invention came during the peak of the Cold War when the Russians were filling their athletes with banned drugs in order to beat the Americans at the Olympic Games (since they couldn't beat them anywhere else) and the importance of winning took such precedence over everything else that no one thought for a minute that there might be ethical problems about using the electrical learning reinforcement stimulator (to give it its official name).

Bizarrely, and despite its moral unpleasantness, the stimulator did actually work. The fear of being zapped with 200 volts of electricity really does sharpen the mind, though of course it can have done very little to instil a love of learning.

COMB-OVER

AMERICA, 1976

Baldness is the one great terror of most young men. If your hair starts falling out when you are fifty you can probably live with it but the prospect of a smooth, domed head at the age of twenty-five or thirty is not a happy one. Apart from anything else a twenty-year-old – as one American inventor pointed out – is less likely to be able to afford a really good hairpiece or wig than a man in his fifties with a good career under his belt.

With that simple fact in mind, the same inventor decided to try to come up with a baldness solution that would work for young, less well-off men, who found themselves follicly challenged.

The inventor was just one in a long line of scientists and inventors who have attempted to find a cure for baldness. Wigs have always been a reasonable standby but they are hugely embarrassing for most people; hair grafts are a more recent option, as is a laborious process of stitching individual artificial hairs into the scalp, but without question the least recommendable of all options for the bald man is the comb-over.

It is extraordinary that so many men still try to get away with this despite a plethora of jokes in the media, but what is even more extraordinary is that an American actually patented a specific type of comb-over. The patent for this still exists and it was accepted on the basis of detailed drawings in 1976. The drawings suggest that the bald man who still has plenty of hair at the back and sides should divide his remaining hair into three sections (one at the back and one each at the sides) before flipping them over on to the top of the head in an elaborate weave.

The most wonderfully eccentric thing about this patented invention is that it would not survive the first gust of wind; nor, given that we all know about how hair grows naturally, would it fool anyone with even remotely good eyesight.

SHEEP COATS

ENGLAND, 1976

Sheep are a valuable commodity despite falling farm prices and worldwide overproduction of lamb and wool. The difficulty of producing wool in Britain is knowing when to shear your sheep – take the fleece off too early and a few heavy downpours or a cold snap may mean high mortality among your flock; leave it a little too late and a sudden hot day can be equally disastrous for sheep still wearing a heavy winter coat.

One inventor saw a clear gap in the market here. He came up with a lightweight sheep coat that would deal with sheep dying from cold as a result of being shorn too early in the year.

Made from jute or canvas, the coats looked pretty much like the sort of coat a dog or a racehorse would wear – fairly thin, but largely waterproof and sufficiently insulating to get a shorn sheep through a few bad days. Odd as it sounds, the sheep coat was actually a good idea until sheep prices got so low that the coats cost more than the sheep they were covering!

HORSE POWER

ENGLAND, 1976

Horse-drawn vehicles for mass transportation may well be a thing of the past but every time there is a fuel crisis – generally when the producers in the Middle East refuse to sell oil to the West or start putting the price up to prohibitive levels – the inventors invariably cast around for alternatives and until hydrogen power becomes a reality they will probably continue to wonder if horses can be utilised in some up-to-date way at least for urban transport where speed is not essential.

In the 1970s the idea of using horses again became a serious possibility for transport in London because a simple calculation revealed that the cost of hay and veterinary bills for a horse was far less than the cost of fuel and servicing on a motor vehicle. Better still, horses had a useful working life of at least fifteen years whereas a car was usually on the scrapheap after six or seven years at most.

But horses are sometimes fickle animals, liable to stop dead or shy at the least surprise, covering the streets with slippery dung and somehow looking out of place in a modern city, so one inventor came up with the idea of a small bus-driven rather than pulled by horses. The bus was very carefully designed with a treadmill placed centrally. Built up around the treadmill was a strong stall in which the horse would have enough room to walk. The passengers sat down along the sides of the inside of the bus, leaving room for the horse in the middle with the driver up in front.

The treadmill was attached to a series of pulleys and gears that led to the wheels at the back. As the horse walked on his treadmill the motion was transported down to the wheels. Up front the driver had a conventional steering wheel and the bus was fitted with electric light, comfortable seats and heaters. When the driver wanted to start he released the handbrake and pulled a lever that was attached to a Heath Robinson apparatus leading to a thick whip-like stick suspended over the horse's rear. When the driver pulled the lever the horse received a tap on the bottom and started to walk. The treadmill turned, drove the wheels and the bus moved off at a stately three miles an hour. When the road was clear the horse would be able to trot and the bus would accelerate to an impressive eight or nine miles an hour, but this was too much both for horse and bus. No horse could be found to trot for more than a few seconds in its stall and even in a very busy city like London a bus that could only travel at three miles an hour was not much use. The prototype was abandoned, but the idea of horse transport in cities has never quite gone away and other inventors are no doubt looking at ways to make it a reality once more.

The ubiquitous Arthur Paul Pedrick conceived a plan for a car driven by a horse attached to an elaborate structure behind the car. The means of acceleration is definitely the best thing about this car – as the driver presses the accelerator pedal the horse feedbag is moved further from the animal's head, making him speed up to reach it!

KEEPING COOL

AMERICA, 1976

Certain inventions remind us that there is some justification for the stereotypical idea of the scientist-inventor as some kind of madman. There are inventions that make no concessions to common sense, for example, or that leave the impartial observer with a feeling of complete bewilderment. Take the sun lounger with a built-in swimming pool. It was conceived during an era in America when anything and everything to do with beach life seemed to find a place in the commercial world. If it had to do with tanning, swimming, surfing or just bumming around on the beach, the scientists and inventors at some big company were bound in the end to invent and market it.

Even allowing for the fashion of the times, it is still difficult to see how this invention ever got off the drawing board. It consisted of a fairly normal-looking if rather luxurious sun-lounger, but instead of being a simple fold-up affair with a padded mattress, it was built with a waterproof mattress and a miniature swimming pool attached.

For those who couldn't afford their own full-size swimming pool in a place where swimming pools were all the rage, this might have made some sense, but according to some reports the swimming-pool lounger was often bought by people who already owned a real swimming pool. It was set up on the side of the main pool, filled with water and used by anyone who'd done enough swimming in the main pool but still liked the idea of lying close – in this case very close – to water.

But what on earth was the point? Experts disagree – but perhaps it was just that in very high temperatures sitting or lying in the middle of your own pool would help you cool as you would enjoy the evaporating water both from the main pool and from your own mini pool.

COLD COMFORT

CANADA, 1977

Waking up in the morning is a serious issue for many people, particularly in the West where people increasingly report intense dislike of their work as well as chronic tiredness. When the alarm goes off in the morning the temptation to roll over and ignore it is, for many, almost irresistible. A smaller number of people regularly oversleep because they simply don't hear the alarm and this can be the case whether the alarm is an intense electronic buzzing or the vast clatter of a big brass bell attached to a traditional alarm clock.

Lots of alternatives have been tried – radios that turn on in the morning are hugely popular – presumably because what they lack in volume they make up for in persistence. Once the radio is on it will stay on till you wake up and turn it off, and the whole process is a little less traumatic than the sort of alarm bell that is so loud it makes you think the house is on fire.

Various inventors have dreamed up a range of ideas to try to improve the waking-up process for those who don't find it easy. These have included Heath-Robinson-like contraptions that operate levers and pulleys to drop things on the sleeper or give him or her a shake; other devices have set off batteries of flashing lights, activated birdsong or whale noises, or even set off a device to pull the duvet off the bed, but a Canadian inventor working in the 1970s must get the prize for the wackiest wake-up invention of all. He conceived a device that involved fitting your bedroom with a pneumatic pump fixed to the wall. This was attached on one side to a hose that led to a freezer unit supplying ultra-cold air. On the other side it was attached to a hose that ran up the wall and over the ceiling to the bed. It was aimed down at the sleeper and when the alarm went off the pump was activated and extremely cold air began blowing on the bed's occupant. In trials, the system worked really well, but volunteers who were woken up in this way said that by the time the discomfort was sufficient to wake them they almost felt they were suffering from hypothermia, with muscles so cold they could barely move anyway.

THE DOG POO SOLUTION

AMERICA, 1978

One of the disadvantages of having a dog – particularly if you live in a city – is the need to carry a selection of plastic bags in your pocket when you take it for walks. The days when dogs could be allowed to foul the pavements without anyone complaining are long gone and hefty fines are now imposed on anyone who fails to clean up after Fido and gets caught.

This has provided a huge opportunity for what one might call the less conventional end of scientific and inventive endeavour. Poop scoops have been designed in all shapes and sizes, from hand-held hoovers to gripping calipers on the end of a stick. Among the maddest of all the devices is a cunningly disguised walking stick fitted with a grabbing and suction action.

Before each dog walk the walking stick owner unscrews the top section of the stick and installs a plastic bag. The stick is then put back together and is ready for action. When the dog does what dogs inevitably do, the walking stick is pointed at the offending item. A trigger is then pressed which activates a battery-powered suction system. This sucks the poo up the stem of the walking stick into the waiting plastic bag, which can later be discreetly removed. The walking stick even included a complex system for sealing each bag and could manage up to ten clear-up operations before it would need to be emptied.

The one drawback was that the walking stick was costly and didn't appeal to younger dog walkers, who hated the idea that people might think they were old enough to need a walking stick.

CRAZY GOLF

AMERICA, 1979

More than most sports, golf is full of passionate players who believe that with enough practice and if only they could perfect their technique, they too could be champions. Instruction manuals for a game that most of us find baffling insist again and again on the importance of the grip and the swing. Arguably these are simply two aspects of the same thing, which is why so many sports scientists have tried to come up with a way to give each and every golfer the edge.

One of the oddest golf inventions is the chin putter and swing guide invented in the 1970s. It looks very much like an elaborate car aerial which is attached to the lower part of the front of the body by a belt. Once in position it is adjusted to suit the individual wearer who, on addressing the ball, aligns himself, his hands and his club with various parts of the swing guide in order to ensure that so far as the swing is concerned he at least starts from the right position. Then, as he actually executes the swing, he has to be careful not to find himself in a huge tangle with half the aerial wrapped round his neck – a risk that may explain why the idea never really took off.

If that sounds cranky in the extreme, what are we to make of the twelve-bore golf club? Only a country obsessed jointly with guns and golf could have put together something as mind-numbingly mad as a golf driver designed to take a twelve-bore shotgun cartridge!

Anyone mad enough to buy the twelve-bore golf club would receive a curious set of instructions explaining its use. The cartridge was loaded into the wooden head of the club (twelve-bore putters were not available in this exploding version) in such a way that the firing pin was positioned in exactly what golfers call the sweet spot. The golfer would know if he'd hit the ball in the 'sweet spot' (proof that his swing was just right) because there would be the most enormous bang. One of the difficulties with the twelve-bore golf club must have been building it in such a way that the pressure gases from the exploding cartridge did not destroy the head of the golf club.

GREENHOUSE HAT

AMERICA, 1980

Hats of various kinds seem to have fascinated inventors over the centuries. The main reason is probably that they provide a perch for all sorts of other devices that can be dangled over the eyes or ears; thus we have umbrella hats, air cooling fan hats, radio-receiving hats, hats fitted with invisible wires and clamps that provide a cheap and instant face-lift, hats fitted with mirror holders and saluting devices.

Among the most interesting is an extraordinary hat patented in the 1980s and designed to bring the benefits of green spaces to the hat-wearer without the need even to get out of bed.

The greenhouse hat is a huge Perspex bubble designed to fit over the head and seal at the neck. Inside the hat at either side of the wearer's temples are shelves on which plants should be placed. While wearing the hat for work or leisure, walking or sitting reading, the wearer can enjoy all the benefits of nature – particularly the oxygen given off by the plants right next to his or her head. For a long time it was thought that this kind of 'fresh' oxygen had special benefits and the theory was that different plants gave off different kinds of oxygen so on Mondays you could wear the hat and enjoy cactus oxygen, while on Tuesdays, you could indulge in spider plant oxygen, and so on. Clips were to be provided to stop the potted plants falling off the shelves should the wearer suddenly swivel his or her head, nod or fall over.

UPHILL SKI

AMERICA, 1980

The biggest problem with skiing is lack of snow. If you happen to really enjoy the thrill of hurtling downhill on two flat bits of wood but you also happen to live in a warm country or somewhere it doesn't snow that often, you have a problem. Most people just accept the fact that skiing is only ever going to be an occasional pleasure for them; others decide that there are more creative solutions. And that explains the invention of snow-making machines. Originally invented to be used at real ski resorts during times when there are huge numbers of skiing tourists and insufficient snow, they are now used in indoor ski centres all over the world. In England, for example, there are at least half a dozen indoor ski-centres where you can ski downhill for upwards of half a mile on real snow – well, at least real snow made by a machine that makes artificial snow. These were a massive improvement on the artificial ski slopes that used to offer a ski surface that seemed to be made from the bristles of a giant toothbrush – bristles that took the skin off your hands if you happened to fall over.

But if the invention of the artificial snow-making machine solved the problems of bare patches on the piste, it did nothing to help those who have snow but no slopes. For these poor souls, an American scientist had the answer: the ski-fan. The ski-fan looks like a rucksack frame with a large propeller attached to it. The propeller sits in the middle of the skier's back and the frame is attached to two handles that run forward on either side of the skier to two hand controls. If you have snow and no hills you can still ski using the hand controls and your skis. The ski-fan simply blows you along the flat. It was also designed for skiers with snow and hills and a dislike of chair lifts and funiculars – with an extra high-powered backpack the skier could actually ski uphill. As with all apparently great inventions, however, there were drawbacks. For a start, one of the pleasures of skiing is escaping from the noise and pollution of the modern world; with a ski-fan strapped to your back you are effectively going to be polluting the pristine slopes with noise and exhaust fumes.

In trials the ski-fan worked really well but the noise of the engine completely outweighed the effortless ability to get back uphill or to ski on the flat, and another imaginative invention bit the dust.

NAPPY NOISES

ENGLAND, 1981

Babies are important. We'd all agree about that, but some people take childcare to absurd lengths. It's not just the insistence on a gigantic, fully armoured four-wheel-drive vehicle to transport the little ones wherever they go. It's also obsessively feeding one's offspring organic low-sugar products, stuffing them with vitamin pills and playing improving music – Mozart and Beethoven – while they sleep. Of course, one can't blame parents for worrying even when their concerns encourage scientists and inventors to come up with increasingly odd devices designed more to remove money from the parent's pocket than to improve life for baby.

Take the electronic automatic cot shaker. This is the Rolls-Royce of cot design. Made from expensive cedar wood (the scent of which is said to make babies feel relaxed), the cot is fitted with a sensitive electronic noise detector. If the baby starts crying, the sensor triggers a gentle vibration to the mattress and the vibrations continue until the baby stops crying. At the same time as the shaking motion is activated, music begins to play from speakers concealed in various places round the cot. These will run through a play list so that the baby who wakes several times in the night doesn't have to listen to the same old tunes. You can also programme the cot music system to repeat various pieces of music randomly, or in any number of pre-specified play lists.

But the musical cot is as nothing compared to the electronic alarm that tells you when the baby's nappy has reached a certain point in the process of saturation. The idea is that the nappy moisture tester eliminates any need to slide a finger down to check the moisture level by hand – a procedure that includes the unpleasant risk that the probing finger meets something a little more substantial than moisture. Of course, most parents quickly get to know when a nappy feels heavy and needs changing but undaunted by this simple fact one inventor outlined detailed proposals for a moisture alarm. According to the patent drawings it was designed as a small sensor that is strapped to the baby's nappy. It slips over the edge of the nappy and one side is actually inside the nappy. When this inside section – which includes the sensor – detects moisture above a certain level, a series of alarm lights begin to flash on and off and the unit emits a bleep.

No one seems to have considered the possibility (actually a certainty) that the baby would be fascinated by this bolt-on device (whether activated or not) and would spend all its time trying to remove it. And apart from the slightly uncomfortable idea of babies' moisture and electricity meeting in the same place, one gets an overwhelming sense with this device that it is, as it were, using a sledgehammer to crack a nut.

SOMETHING FISHY

AMERICA, 1982

For some people, fish in a tank are to be ranked among the most boring things on earth. For others an aquarium is the closest thing to heaven on earth. The fact that tropical tanks are hugely popular can be judged from the number of shops that sell fish and the equipment necessary to keep them alive at home. Then there are the clubs and societies devoted to keeping fish and the numerous magazines that offer features and news about all things fishy. For the rest of us the whole thing is baffling. But even within the ranks of a pastime that many of us find incomprehensible there are those who consider some fish-keeping ideas beyond the pale.

One that never really took off was the tropical fish tank bath. This was a full-size bath made from a strong, completely transparent plastic or (in a more expensive model) bullet-proof glass. Within the main bath and smaller than it by at least a foot was another bath, also made from completely transparent plastic. The smaller bath was fixed to the large bath only at the bottom. This meant in effect that there was a foot-wide gap running all round the inner bath and down the side. The idea of the fish bath was that this outer corridor, as it were, should be filled with water and stocked with fish. All the necessary fish equipment – oxygenators, clumps of weed, shoals of gravel and so on would be placed in the narrow outer tank and plenty of colourful fish introduced.

When the fish owner needed a bath he or she would use the inner bath and be able to lie there surrounded by his favourite fish, all apparently swimming perfectly naturally in their narrow tank. For the fish enthusiast it would be like swimming in some tropical ocean down among the brightly coloured denizens of the deep.

NABBED BY NET

ENGLAND, 1982

The trouble with insurance is that it's hard to put a value on sentimentality. Family heirlooms may not be worth a fortune but it is far more upsetting to lose them than to lose, say, the latest hi-tech DVD player or television.

The same is true of antiques – you can have them valued and pay for the appropriate level of insurance but if they are stolen they may not be replaceable since replacements cannot be found as easily as new items of furniture, pictures etc. A British inventor concerned at the risk to his own antiques decided that the solution to the security problem for antiques wasn't to have an elaborate alarm system fitted to his house but rather to fit an alarm to every single antique in the house.

For chairs, tables, pictures and virtually every other conceivable movable antique he created an inexpensive alarm that would be triggered by a tilt mechanism. In short, if there was any attempt to lift or move the antique the alarm would go off. The difficulty was that the householder had to go round every night turning each and every alarm on and then repeat the procedure in reverse in the morning. Forget one alarm and the least movement meant a disturbed night and the possible arrival of the police.

A more intriguing invention involved half a dozen six-inch-diameter holes being drilled in the walls of each room in the house. One hole would be drilled in the centre of the wall of each of the main downstairs rooms and then packed with a special explosive and a tightly compacted super-strong fishing net. When the alarm was set at night, or when the occupants of the house were on holiday, any movement detected by infrared beams in any of the rooms immediately detonated the explosive nets, which shot out across the room, completely entangling any would-be robber.

The invention was excellent from a technical point of view and was based roughly on the principle by which car air bags work. The inventor's huge enthusiasm for his ideas did, however, make him forget one detail: when the nets shot across the room they entangled the burglar *and* all the furniture, which meant that the burglar was partly protected and given a few moments to untangle himself he would certainly be able to escape. Tests in an empty room revealed that the net would tangle a burglar for at least twenty minutes – long enough for the police to arrive and make an arrest. But then why would a burglar bother with an empty room in the first place?

HOUSE AND HOME

ENGLAND, 1982

In England houses have been built from the oddest materials. Cobb houses, which still exist in large numbers, were built from lumps of clay pounded by horses' hooves. You put the clay and plenty of chopped straw into a trench and then persuade a carthorse to walk up and down on it for a few days; the resulting blocks are as hard as iron and can be cut to fit.

Other early houses with a basic timber framing had infilling made from sticks, twigs and horse manure or mud and in Scotland and Ireland some roofs were made from turf with the grass still growing on it until well into the twentieth century.

More recently eco-houses have been built from straw bales – when compacted, straw is strong, stable and a superb insulator.

Among the craziest inventions was the house built as an experiment in Northamptonshire – from old books. Like straw, paper is a good insulator, and books can be stacked neatly to form solid walls. The only difficulty was that the house was built from untreated books; had the book pages been glued together and the whole thing treated with plasticiser, all might have been well but the house of books was actually more like a house of cards and it swelled badly as soon as it rained and was prone to large movements as the weight of the floors caused the pages within the books to move.

Within weeks, the book house was given up as an impossibility but the books weren't wasted – they were densely compressed to create useful, long-lasting logs.

BUTTER STICK

JAPAN, 1982

More than any other nation on earth, the Japanese love gadgets. Not only do they love them but they also go to great lengths continually to think up new ones. The Land of the Rising Sun gave us the Walkman, for example, and much of the technology for later inventions – MP3s, DVD games and so on – is also Japanese. But Japanese scientists have taken the idea of inventing gadgets a step further and they now deliberately try to invent elaborate and beautifully made gadgets to do a very specific (and often virtually pointless) task. The fact that they fail in this attempt is nothing to be ashamed of, as elegance is the point of the exercise. The Japanese have even invented a word for these stylish but ultimately useless bits of scientific inventiveness – they are known as *chindogu*.

Among the truly great *chindogu* inventions are slippers for cats and dogs that double as floor dusters; a rack for drying wet clothes that attaches to the top of a car; and a small fan that is attached to a pair of chopsticks. The idea here is that as you start to eat you turn the fan on and then as you lift the food to your mouth it is cooled – in transit as it were – by the whirling fan!

Despite the fact that the real aim of these *chindogu* inventions is that they should seem plausible rather than actually be of any use, occasionally the system breaks down and, horror of horrors, a *chindogu* is taken up by a manufacturer and considered as a real, practical possibility. This is what happened with the butter stick.

The butter stick was designed pretty much like a deodorant or glue stick with a solid mass of butter concealed in a plastic tube. When you needed some butter you took the lid off the tube and pushed the plunger at the other end of the tube until the butter appeared. All you had to do then was rub the thing over your slice of toast, saving the need for knives to do the spreading!

There are believed to have been some trials to see if the public liked the idea of the butter stick and there's no doubt that had the trials been favourable the company would have put the new device into production. The difficulty seems to have been that the butter stick was too reminiscent of deodorant sticks and appears to have put people off, but it is probably one of those ideas that will eventually be produced – perhaps when people in the developed world are even busier than they are already and the prospect of being able to enjoy a piece of toast without dirtying a knife seems a useful time-saver rather than the product of an imaginative inventor's over-active brain.

TOILET HUMOUR

AMERICA, 1982

As in so many spheres of life, tragedy and comedy are often close cousins in the world of scientific invention. We all know that skyscrapers, though spectacular to look at, maybe even to live and work in, are terrifying if anything goes wrong and you happen to be inside and need to be evacuated quickly. If there's a fire in a tall building and you can't escape you are in big trouble and of course trying to get out of a fifty-storey building is far more difficult than running out through the front door of your average suburban semi.

Numerous ideas have been canvassed for coping with fires in tall buildings, some more realistic than others. Sprinkler systems work well as do foam smotherers. Slightly less conventional ideas include a proposal to leave a giant blanket rolled up to the top of the outside of every tower block so that in the event of fire the blanket can be released to envelop the whole building and, of course, the fire itself.

Perhaps the strangest idea of all was a complex piece of equipment patented by an American scientist and inventor in the early 1980s. It was designed to enable anyone stuck in a building on fire (but especially a tall building) to survive inside the building for much longer than would otherwise be the case, until help arrived.

The device was known (and may still be, for all anyone knows) as a toilet snorkel. It involved a helmet and mask fitted with an air hose. When the going got really tough you were supposed to rush into the nearest lavatory, push the air hose down through the water in the bottom of the loo and reach fresh air on the other side of the water barrier.

The logic of the device was impeccable. However much smoke there might be swirling around you it would not be able to breach the water barrier at the bottom of the loo. In theory then, if you are able to access the air on the other side of that watery smoke barrier you should increase your chances of survival because most people who die in fires die from smoke inhalation or asphyxiation caused by lack of oxygen.

Sadly, our intrepid inventor forgot one important thing – the whole point of the water barrier in a lavatory is to keep the noxious gases from the sewage system from seeping back up into your house or office. Trying to breathe these gases might not be quite as damaging as breathing smoke fumes but it would be extremely unpleasant and might well kill you anyway!

LEVITATION AT LAST!

AMERICA, 1983

For centuries it was a basic rule of science that levitation was impossible. Other than powered flight or hot-air ballooning, there was no way that a heavier-than-air object could be kept aloft without the aid of physical props of some sort.

During the eighteenth and nineteenth centuries a few scientists thought basic problems could be overcome using magnets, but one person who did more than most to see if magnets really would allow levitation, the Reverend Samuel Earnshaw, declared in 1842 that it just couldn't be done using ordinary magnets (with electromagnets it was certainly possible). Earnshaw's Theorem, as it became known, was accepted as fact until well into the twentieth century, but this didn't stop amateur and professional scientists tinkering with the idea.

But in the mid 1990s an amateur American scientist created a device that did seem to overturn Earnshaw's Theorem and allow genuine levitation using ordinary magnets. What is even more remarkable is that the Levitron, as the device was known, had been made using just a few magnets and plastic pieces.

The Levitron is now available as a kind of highly sophisticated toy – but despite being a toy it does still represent a strange and remarkable invention: an invention that reverses traditional scientific wisdom.

So what exactly is a Levitron? It consists of a base fitted with a magnet and an extremely lightweight spinning top which is also fitted with a magnet. The base must be set up so that it is absolutely flat and the levitating magnet has to be very precisely adjusted; but carefully set up, the Levitron really does allow levitation to take place.

The Levitron was patented by Bill and Edward Hones of Seattle but the idea was the brainchild of another American, one Roy Harrigan, an eccentric but hugely talented amateur scientist from Vermont. His levitation device had been patented as early as 1983.

The Levitron has one weakness – the magnet will stay in the air only while it spins and friction caused by spinning in the air means that levitation lasts for a couple of minutes at most. Within a vacuum the top will stay in the air for several minutes but even then it will eventually fall for reasons not yet entirely understood.

But the Levitron is still a most bizarre invention. It looks like a metal dish a few inches in diameter on legs. The spinning top is pushed up into the air away from the base plate because its magnetic north pole is facing down while the base plate magnet has its north pole facing upwards. The two magnet forces repel each other and at the right height the magnetic repulsion force cancels out the force of gravity. The point where the top hovers is the point of equilibrium.

The spinning motion that has to be imparted to the top in order for the Levitron to work is vital because the base magnet exerts a slight side pull which would otherwise turn the top over, thus destroying the north–north opposition of the poles and bringing the top crashing to earth. With spin the top acts like a gyroscope, but it still has to be set in motion with precision.

The height the top spins is always 1.25–1.75 inches above the base plate because this is where the point of equilibrium of the forces will always be achieved. Tiny washers have to be taken away or added to the spinning top according to the temperature and humidity in order to achieve the exact weight required and the top itself weighs less than a postage stamp!

All the effort to produce a scientific invention that so far has little practical application may seem wasted, but scientists enjoy proving each other wrong and the dismissal of Earnshaw's Theorem is certainly a huge achievement.

CHILD'S PLAY

ENGLAND, 1984

Children seem to have almost unlimited energy. They rush about for hours on end and many parents must have often wondered if there wasn't a really useful way to harness all that energy. At least one inventor clearly thought so when he created the tricycle lawn cutter. His theory was that the cutter could be made in a variety of sizes for children and for adults. It was simply an extra-sturdy tricycle fitted with extra-thick tyres with the sort of tread you would expect on a mountain bike to give added grip. A series of interlocking blades – like those used on old-fashioned push mowers – was attached around the back axle of the tricycle cutter so that when it moved the blades turned. Above the cutter was fitted a metal tray to catch the cuttings and a simple device allowed the blades to be raised or lowered. The tricycle grass cutter had the additional advantage that it provided good exercise. The disadvantage was that even on a smooth lawn it would have been hard to get the bike moving and the least dampness would have caused skidding which would have spoiled the lawn anyway. Extra weight added to the bike to improve its grip would have made it even harder to pedal, and so on.

The real problem with the lawnmower was that its motive force – pedalling – relies on the friction between ground and wheel to get it moving. Inevitably that is hard on the surface involved. By contrast, a conventional powered mower, even of the large sit-on variety, has wheels that are driven from within, and adhesion to the ground is less of an issue.

MICHELIN MAN

ENGLAND, 1984

Safety is no laughing matter when it comes to road travel. Driving a car is vastly more dangerous than flying yet hardly anyone has an aversion to travelling by car. With aeroplanes it's entirely different and tens of thousands of people are so terrified (despite the statistical evidence that flying is incredibly safe) that they simply refuse ever to get aboard a plane and would rather spend a couple of weeks travelling overland to their favourite destination.

Back on the ground, in Britain alone more than 6,000 people a year are killed or seriously injured in road accidents – that includes cars (dangerous), motorbikes (very dangerous) and cycles (extremely dangerous).

Dozens of organisations around the world have spent millions of pounds trying to make road travel safer and there is no doubt that the situation is a lot better today than it was twenty or thirty years ago. For a start, drink-driving in Britain has been outlawed and there are increasingly severe penalties for those who drive dangerously or without due care and attention. Car brakes are much improved, seat belts are compulsory, helmets must be worn by motorcyclists and cars are sturdier.

But the scientists and inventors are never complacent about these things and neither amateurs nor professionals – the latter usually employed by the car and bike companies themselves – can rest on their laurels.

One highly creative English inventor devised a safety suit for motorcyclists that sounds excellent in principle and may well have worked in practice. His idea, developed in the early 1980s, was for a suit that would inflate in a fraction of a second at the first sign of trouble. Air bags were already being fitted to American cars at this stage but they couldn't – for obvious reasons – be made to work on motorbikes so our intrepid inventor simply adapted the air-bag concept.

The inflatable suit was made in a tough neoprene material fitted with small gas cylinders at key points within the fabric. Any sudden deceleration or collision would trigger tiny explosive charges that would ignite the bags before the poor motorcyclist hit the ground, the other car or a tree. In short, the motorcyclist would be turned into a huge bouncy ball and any impact would be absorbed by the ball rather than the man in the middle of it.

The only fly in the ointment was the fact that the suit was hugely expensive and it was very difficult to adjust it so there was no danger of the suit accidentally inflating while everything was proceeding normally. The insurance industry would not take kindly to the risk of numerous motorcyclists suddenly ballooning up off their bikes and causing as many accidents as the suit was designed to save them from.

SPIDER LADDER

ENGLAND, 1985

An invention that seems odd when it is first registered at the patent office may never get off the drawing board simply because it is an invention before (sometimes after!) its time; other equally weird inventions can suddenly find that they really are not so absurd after all. Catseyes – little reflectors built into the centre line of virtually every road in Britain – are a case in point.

Another classic example which sounds unbelievable at the outset has actually meant high earnings for the entrepreneurs who originally made and patented it.

The invention is the Bath Spider Ladder.

The inventors clearly thought they could tap into the popular imagination on two levels. First, animal-lovers who hate washing spiders trapped in their baths down the sink, and second, people who dread spiders so much that they cannot bear to go near a bath with a spider in it let alone deal with such a spider. The spider ladder gets them off the hook.

A narrow plastic ladder with a curved end, the spider rescue ladder is simply attached to the inside of the bath in any convenient spot. Whenever a spider falls in it will eventually find the ladder and make good its escape. For the arachnophobic the ladder is a godsend because it can be left permanently in position, reducing to an absolute minimum the chances of finding a lost spider in the bottom of the bath.

Made originally perhaps as a bit of a jokey gimmick, the spider ladder has sold all over the world and tens of thousands of spiders are only with us today because of the brilliant conception and design of what was once seen as a preposterous and useless invention.

CUTTING EDGE

SOUTH AFRICA, 1985

Getting a haircut isn't the most arduous thing in the world yet numerous inventors have tried to come up with systems to enable us to avoid that regular trip to the barbershop. Among the oddest has to be the machine patented by South African Jan Louw.

Mr Louw invented a curious device that harnesses the power of your vacuum cleaner in such a way that you can cut and style your own hair without inadvertently lopping off one or other of your ears.

So confident was he that he patented the device in no fewer than eleven European countries! The hair-cutter looks rather like a hairdryer but it is attached via a special nozzle to the hose of a vacuum cleaner. When you turn the vacuum cleaner on, the suction draws air in and over the blades of a small turbine; the turbine has enough power to drive a conventional-looking pair of hair-cutters, which work by running one blade rapidly back and forth over a fixed blade.

Anyone who has a regular – and now very fashionable – razor cut will know how easy it is to use this kind of hair-clipping device. You simply run it over the head and all the hair vanishes in an instant. Where Mr Louw's cutter differs from conventionally driven electric models is that it also uses the suction to pull the hair into the cutting blades, using nozzles of various lengths to vary the length of the cut. All the hair that is cut is neatly sucked into the dust bag of the vacuum cleaner.

Now, on the face of it, this probably doesn't sound like a strange invention at all – after all, it's highly practical, efficient and would probably have been relatively cheap to produce and distribute. The most inexplicable aspect of the vacuum cutter is the fact that the inventor assumed people would prefer a device that ran off a vacuum cleaner to one that ran simply by plugging it into a socket. If you already have a device that does its job effectively why invent another device that does the same job in a more elaborate and outlandish fashion?

SKI BRAKES

AMERICA, 1985

Anyone who has learned to ski will remember with a shudder of horror the sensation of massive acceleration when you lose control. When you're learning to ski the sudden prospect of a steep slope causes an instinctive tendency to lean back, but if you do that on skis you lose more control and your speed increases still further, which makes you lean away from the horror still further and so on till you fall in a crumpled heap. Once you get the hang of it you can do the counter-intuitive thing and lean into the slope when you begin to accelerate – this gives you more control and ironically lessens the chance of a runaway disaster.

But at least one inventor decided that learning to ski properly wasn't the only way to avoid uncontrolled and terrifying mountain descents. He believed that skis were just another kind of vehicle and should therefore be fitted with a braking mechanism. He came up with the ski sail. This was basically just a plastic sheet that stretched between the skier's poles. Any unforeseen acceleration would be blocked by the wind resistance against the sheet.

The inventor insisted that the real pleasure of using the brake was that descents were always sedate and it was a new pleasure to feel the pressure of the wind against the sail. An English inventor came up with a similar device at about the same time, based on the successful deployment of air bags in cars. When the car crashes the air bag inflates in a fraction of a second to protect the driver and passengers from injury. On skis a similar effect could be achieved – or so the inventor thought – by creating a pouch on the front chest area of every skiing jacket sold to a beginner.

While everything on the slopes was going well the jacket looked perfectly ordinary, but if the skier got into difficulties by speeding up and perhaps even heading for a disastrous crash with a tree or telegraph pole he or she would pull on a special toggle that would instantly inflate what looked like nothing so much as an open umbrella. The hollow of the open umbrella would face into the wind, creating a strong braking effect. In trials the difficulty was that the skier tended to be hurled backwards by the suddenly opening brake. This led to a decision to try a parachute released from the back of the jacket, but this was little better: skiers who'd lost control and were hurtling down the mountain found that, at the touch of the toggle, they were yanked backwards into the air and unceremoniously dumped on the piste.

HAT TRICK

ENGLAND, 1987

Camping is one of those pastimes that attract inventions as a candle attracts moths. This has a great deal to do with the obsession with making things ever lighter and ever smaller. Many inventions – using ultra-lightweight modern materials – have been hugely successful, for example, the tent that relies on springy fibreglass poles to stay upright rather than on old-fashioned and heavy steel pegs driven into the ground.

Yet others are not quite so sound and among the most delightfully amusing has to be the hat that turns into a tent. The idea behind this was that given the often poor British summer weather a bigger than average hat would not be a problem – indeed, it might be welcomed. A big hat would also allow scope to use the extra material involved to create something even more useful than a hat. Thus was born the one-man tent that folded away to create a large-brimmed hat.

The tent was basically a waterproof tube made of a high-strength and lightweight nylon that had been impregnated with a waterproof silicone. Six ultra-lightweight fibreglass poles came with the hat, although they had to be carried separately. When you decided to pitch camp you took off your hat, unfolded the material, added the thin, springy poles and your tent was ready. The tent was cleverly fitted with permanently marked fold lines so that it could be folded back into a hat after use.

Wearing your tent on your head had other advantages, too – it meant the weight of the tent was directly above the man wearing the hat (rather than dragging backwards like a rucksack), but the disadvantage was that anyone wearing the tent hat tended to look rather like one of those traditionally dressed West African women on their way to market with a large bundle of goods piled on top of their heads. It was perhaps for this reason that the tent hat never really got off the drawing board.

UNDERWATER GOLF

AMERICA, 1988

Someone once claimed that without golf America would not be America at all but a sort of banana republic, a place unworthy of God's chosen people or in fact any people at all. The man in question was a fanatic but the popularity of the sport is not in question – from coast to coast golf is a sport that unites all American men who like to wear loud checks and escape their wives at the weekends.

Knowing this, numerous ingenious inventors have tried to find ways to further enhance the lives of golfing enthusiasts. The famous (or infamous) British inventor Arthur Paul Pedrick, for example, came up with a golf ball with small magnetic wings that were activated by centrifugal force to reduce spinning if the ball was incorrectly struck. Others have designed electronic tees that record the speed and angle of departure of the ball. Among the best and arguably the maddest of golf-related inventions is that conceived by an American who felt sorry for golfers who had to go on beach holidays every now and then with their families. His solution to the hell of being away from the golf course was to come up with an underwater golf swing training device.

What may sound rather sophisticated was actually just a square of plastic or metal on the end of a handle the length of a golf club. An adjusting screw allowed holes in the square blade to be opened and closed to increase or decrease resistance. No one has yet quite made out why being able to reduce or increase water resistance should be good training for swinging a golf club through far less dense air, but be that as it may, the illustration for the patent for this unparalleled contraption shows a keen golfer standing up to his neck in water swinging back and forth at an imaginary ball.

There is nothing in the patent that explains to the golf addict what he or she should do if the water is murky and the imaginary golf ball can't be swung at (because of poor visibility) by the plate on a stick!

LEARNING GLOVE

AMERICA, 1989

Devices to help children and adults learn are a staple of the inventor's repertoire. From pop-up books at the simplest end of the spectrum to elaborate electronic learning systems based on mind and intelligence theories, the world of learning is a rich source of material.

Among the maddest learning devices ever made was a range of gloves dreamed up by an American inventor in the late 1980s and early 1990s. Instead of picking a pair of gloves on the basis of their colour or the material from which they were made you chose your gloves according to what you wanted to learn. Children's gloves were available covered in Spanish lessons or French grammar; adults could buy gloves that taught them the intricate secrets of particle physics or philosophy.

The theory was that the odd bits of information you could concentrate on while waiting at the bus stop or strolling through the park were more likely to stay in your head than reams of information you might go through in a sleep-inducing library.

Scientific evidence about how the mind works does suggest that this approach has something to recommend it – the mind is better, generally speaking, at picking up small amounts of information that are studied repeatedly and in the case of the gloves the idea was that you would have to look at them on a long walk because the words or formulae would be there every time you lifted a hand to your face or took the gloves off or put them on. Their effectiveness was based on the idea of reinforcement, but as is so often the case it was hard to get people to part with money for something that seemed – to say the least – eccentric. But at least one internet website has introduced something rather similar to the glove idea: register with them and every day they send you a new word or phrase in a foreign language. In this way you can build up a wealth of knowledge by the relatively easy business of remembering a small amount every day.

BACKPACK BED

AMERICA, 1990

When long-distance walking and camping began to grow in popularity in the 1960s and 1970s, enthusiasts at first had to put up with the fact that there really was very little specialist equipment available for their favourite pastime. Rucksacks tended to be old heavy canvas Army surplus things with few pockets and a single large compartment that allowed little serious packing – your only option was to throw everything in a heap. Worse, the canvas was not waterproof. The same was true of tents – where they were available they were made from heavy canvas and they usually came without groundsheets. The problem was that tents were still being made by companies who had been making them – largely for the Army – since the Crimean War. Campaigns in hot countries gave the tent without a groundsheet some advantages – it was cooler, for example – but hiking across the Pennines with such a tent meant wet, miserable nights. Boots were also primitive at this time, to say the least – Army-issue boots were the only real option and since they were designed at least partly to toughen young recruits up by making them endure agonizing blisters, they left ramblers and walkers scratching their heads for a solution. Walking long distances was a fairly new pastime, and modern lightweight materials were in their infancy, but within ten years the range of camping and walking equipment being made by specialist firms to meet the needs of the new breed of walker had exploded exponentially. But among a huge spectrum of excellent boots, tents and rucksacks, there were odd items that might at best be described as eccentric.

Making cold-weather walking gear is a real science, particularly at the high end of the market where the user might well be climbing Everest and the quality of product could mean the difference between life and death.

A remarkably complex invention from the early 1990s might at first have seemed the ultimate in outdoors design – the combination backpack, camp bed and tent on the face of it cleverly combined three of the heaviest (yet most important) outdoors items in one, thus making savings on space, weight and overall efficiency. However, the patent drawings for the combo bed reveal something that looks as if it might well close up suddenly while you were sleeping and trap you in a mass of wires, struts, folds and rivets. The other difficulty is that each of the three items in the combination had to sacrifice something in order to allow the transformation into the other two items. Thus the bed would not have been very comfortable while the rucksack lacked the carrying efficiency of simpler rucksacks. The tent, too, somehow lacked the lightweight precision and weatherproof qualities of tents designed to be just that.

The irony is that despite the undoubted talents of the inventor, the combination bed, rucksack and tent couldn't compete with top-quality items sold separately. In fact, having the three items as separates was actually far more efficient than having them combined and without the need for an elaborate amount of design technology aimed solely at making the three work together rather than making them work really hard for the user.

MILK GUN

AMERICA, 1990

It would probably be unkind and certainly unfair to accuse all Americans of being crazy about guns. But there is no doubt that the spirit of the Old West still lingers, particularly in midwestern states where many people still think Nicaragua is in Europe and that any reduction in the number of Kalashnikovs kept at home could result in a renewed spate of Indian uprisings.

Only this extraordinary passion for the right to bear arms – as the American Constitution puts it – could explain an outlandish baby-feeding device invented in the 1990s. The baby milk gun is a sophisticated piece of equipment with dozens of complex moving parts. It also looks like a real gun.

The idea behind it was to produce a method of feeding baby that was both quick and effective but which also avoided any mess. And there is no doubt that with its carefully calibrated nozzle and skilfully designed delivery system the baby-milk gun does just that; but what the inventor seems to have completely failed to appreciate is the effect the baby-milk gun would have if used in public.

Imagine the consternation of shoppers or fellow diners in a restaurant when a mother with a cute-looking baby pulls out what looks like a revolver and sticks it in baby's mouth. As in so many cases, the passion for invention seems to have blinded the inventor to some of the more obvious issues.

AIR HEAD

AMERICA, 1990

The world's biggest-selling drug – now outstripping even aspirin – is Viagra. Hardly surprising really when one remembers that it's a drug that solves an age-old and highly sensitive problem that apparently afflicts a large number of men. If the equivalent to Viagra is ever found for baldness, the discoverer or inventor is likely to become a millionaire and win a Nobel prize.

Novel inventions for hairlessness have included bizarre mixtures of guano and honey, electrical shocks to the scalp, the liberal application of horse dung and much else besides. And all to no avail. As late as 1990 devices with no basis in science were still being invented and then touted as genuine cures for a condition that has baffled the best scientific minds for centuries.

An inflatable headband and cap was invented in America on the basis of fairly flimsy research that suggested hair might be encouraged to grow again if the scalp was stimulated in the right way. Complex and expensive and fitted with a pump and valves to regulate the pressure, the scalp stimulator fitted tightly over the head. It was then attached via a high-pressure hose to a large canister of helium gas. As the valve on the gas was released the cap filled with helium, which pressed down on the bald scalp. If used every day for an hour or so it was hoped that the helium combined with pressure would soon have a veritable jungle of hair growing even on the worst case of male-pattern baldness.

But the inventors hadn't done their homework for by 1990 when the scalp stimulator was invented the science of genetics could have told them not to waste their time: baldness is written into the genes (or not, if you're lucky) and once the hair-producing cells switch off no amount of electrical or gas stimulation is going to switch them on again.

The only hope for those who are follically challenged lies in the continued rapid advances of genetic research for it may be that scientists will one day be able to turn back on the genes that create a fine head of hair in youth and thereby avoid a shining bald pate in middle and old age.

UMBRELLA TO THE RESCUE

CANADA, 1990

Many great inventions (and an equal number of really wacky ones) are based on existing inventions. Take the umbrella – it's been with us for centuries and is a simple and effective way to avoid getting soaked. But at least one inventor saw far more potential. Canadian Gary Kassbaum decided that the umbrella, with a few modifications, could be used as a life-saving device in accidents where tankers – road tankers or huge ships carrying oil and other flammable materials – find they have been holed.

The new life-saving umbrella was, to all intents and purposes, indistinguishable from a real one – the difference was that the safety umbrella was made from far stronger materials. With steel struts holding the canopy open, the Kassbaum brolly also had a ripcord running down the metal shaft. In an emergency the umbrella would be pushed through the hole in the tanker and once inside, the operator would pull the ripcord, thus opening the umbrella within the body of the tanker. The enormous pressure of whatever flammable liquid remained in the tanker would act on the umbrella's large surface area, forcing it against the hole at enough pressure to prevent any further leakage. On the outside the umbrella handle was designed to clamp to the tanker's side, thus holding the whole thing in position.

That at least was the theory – the practical difficulty might be getting close enough to the holed tanker to push the umbrella through the hole without being washed away in a tide of oil or liquid gas!

WATCH THE DOGGIE

AMERICA, 1991

The prize for the ultimate silly invention might well go to the dog watch invented in 1991 in America. It's hard to imagine why it might be even remotely useful, but the inventor clearly knew a thing or two about the American attitude to pets. The watch was designed to give the dog's owner a sense of how time is passing for Fido. How did it do this? Well, it worked on the oft-quoted principle that every year of a dog's life is equivalent to seven years of a human life. The dog watch is a fully operating watch but for every minute of human time the dog watch gives seven minutes. At any given moment the dog owner can then estimate how old their dog is and the rate at which he or she is ageing.

One problem with the dog watch is that dogs – ungrateful creatures – seemed to be very unhappy about wearing it and dog time tended to become no time at all once the watch had been given a serious chewing in an attempt to remove it.

WRIST ACTION

AMERICA, 1992

Most hospitals in England, Wales and Scotland are busiest on Friday and Saturday evenings because that's when people between eighteen and twenty-five go out, get drunk and either fall over or beat each other up. The binge-drinking phenomenon has baffled social scientists and politicians for the past decade and more and various theories have been put forward to explain the peculiarly British attitude to alcohol. The generally accepted explanation is that the British are addicted to the idea of celebration – when the week's work has finished they head off to pubs and clubs to get outrageously drunk because a huge celebration matches their mood of elation.

We also seem to think that there is something inherently heroic, even artistic, about getting supremely drunk. We don't boast that we had a great night out but that we were completely tipsy, pickled, plastered, even slaughtered – the vast number of terms for being gloriously blindly drunk is a reminder of just how central intoxication is to youth culture in Britain. But as the politicians and other more sober-minded elders put their minds together to try to get to grips with the social costs of drunkenness, the scientists and inventors see potential for their activities and among the more amusing inventions designed to at least tell the drinker that he or she may be going just a little too far is the binge-drinking bracelet.

Invented in 1992, the bracelet contains a simple counting mechanism and an alarm. It can be set before you go drinking to allow you four, five or more drinks but will then roar into action if you exceed the pre-set number of drinks. How does it know when you've exceeded your limit? Easy – it registers a drink each time you lift your hand to your mouth. That does raise the slight difficulty that you have to calculate in advance how many sips you will need to finish your pint or two pints, but once you've made the calculation that bracelet won't let you forget the deal you made with it when you were still sober. For some extraordinary reason the bracelet was an instant flop and though it remains as a registered patent it was never put into production.

ANGRY ASHTRAY

AMERICA, 1992

A Taiwanese inventor concerned at the huge loss of life caused by the smoking of cigarettes decided that warnings in newspapers and on television and on the radio were not enough. He decided that smokers really needed to hear a warning just as they were about to indulge their dangerous habit, so he invented the angry ashtray.

The ashtray, which looks pretty conventional, has a side tray that slides in and out for matches, but the bottom of the sliding drawer is highly sensitive to movement and when a match is removed by a smoker a tape recording is activated and transmitted loudly through a tiny microphone – the noise of loud cancerous coughing fills the air.

A later development of the ashtray includes a larger separate drawer where the smoker can store up to one hundred cigarettes. But this drawer too is fitted with a highly sensitive mechanism. When there is any attempt to remove a cigarette a pre-recorded tape is again activated only this time a stern voice booms out, warning about the dangers of cancer, heart disease and strokes.

Anyone interested in the ashtray could choose one with a relatively mild rebuke or one that is downright abusive – the ashtray, in other words, shouts at the smoker telling him or her they are an idiot who will die a slow and very painful death.

Surprisingly, the angry ashtray was not a commercial success!

SINGING CONDOM

AMERICA, 1992

Almost everyone wants to have sex but no one wants to catch a sexually transmitted disease. The problem is almost as old as human history and the invention of the condom was one of the great breakthroughs. Condoms, though, are by no means a modern invention. The eighteenth-century writer James Boswell describes in his diaries having sex with a girl in 'armour', by which he certainly meant a primitive form of condom probably made from sheep's guts and tied on with silk tapes!

In 1992 an American inventor patented an interesting variation on the condom principle. His idea was to fit each condom with a small pressure-sensitive pad which also contained a microchip, a tiny battery and transmitter. At each thrust the pressure pad would activate the microchip, which would then play one of half a dozen electronic tunes – rather like those you hear on a mobile phone. Thus for certain kinds of sex one could have romantic tunes, for others perhaps a triumphal march or the *Ride of the Valkyries*!

Even more interesting was the brilliant invention of the full body condom. Made from a fine latex and skin tight, the full body condom is fitted with breathing holes (which actually seem to be complex breathing tubes), high-quality eyepieces (yes, even the eyes are covered) and – would you believe it – a condom. So, in essence, this is a condom fitted with a condom.

The full body condom was seen by its French inventor as having wider applications than absolutely safe sex – he also saw it as a way to protect against all types of viral and bacterial infection. The difficulty is that the discomfort of wearing the body condom is probably greater than the discomfort one would be caused by most common infections.

FLY TRAP

AMERICA, 1993

Those fearsome-looking blue Electrocutes that hang in food shops around the world are probably the ultimate insect zappers. They look rather like old-fashioned electric fires but with blue bars rather than red. The clever thing about them is that they don't just zap anything that passes by; the light they emit is specially designed to attract insects from some distance. In fact it is believed that the light somehow has an effect on the insect that parallels the effect of hormone attractants in the opposite sex.

But Electrocutes don't look great in the home where traditionally we've used sticky flypapers or fly swats or – for the green fingered – Venus Fly Traps. Plant fly traps are good but difficult to keep alive for the amateur, which is why an American inventor came up with an artificial fly-eating plant.

This offbeat idea actually had much to recommend it, at least for those who didn't object to large basinfuls of plastic blooms all over the house. The artificial Venus Fly Trap was basically a bunch of hollow-stemmed plastic flowers in a pot filled with insect attractant. This had to be a chemical insect attractant because jam or honey would be messy to deal with and would go off quickly anyway. The inside walls of the hollow stem of each plant were lined with fine downward-pointing hairs and the only way to reach the insect attractant was down through the plastic bloom on each plant and then down through the stem. Once it had reached the insect attractant, the fly, wasp or other insect would find that the downward-pointing hairs in the flower stems prevented its escape. In other words, this was a hotel where you could check in but you couldn't check out.

It is difficult to know quite why this idea quickly reached the inventions junkyard but that's exactly where it ended up. Perhaps by 1993 artificial flowers had completely lost whatever minimal popularity they ever had; or maybe it just seemed crazy to buy an artificial version of a real plant when both would do exactly the same job.

LIGHT UP THE LOO

AMERICA, 1993

Some inventions are so delightfully dotty that we are inclined to forgive their inventors for coming up with something that from a practical point of view is likely to be an expensive waste of time. But then many inventions succeed despite being rather mad because it becomes fashionable to own them. Their very uselessness is their attraction. A Japanese invention is a case in point: it was designed to be attached to a loo roll and counted the number of sheets being used. It could be set so that an alarm went off if the user took more than five, ten or fifteen sheets in one go. Or take the clock that runs backwards – it does tell the time but you have to think about it.

In 1993 an inventor became concerned when he noticed that older people tend to have to get up once, twice or even three times in the night to go to the loo. He couldn't do anything about this directly – it's just a sign of ageing – but he thought that a device that made the experience of getting up more pleasurable or at least less annoying might be a viable proposition.

He did some research and discovered that the thing that really annoyed older people when they had to go to the loo in the night was the fact that they had to turn the light on. The bright light automatically woke them up and made it difficult to get back to sleep, but without turning the light on there was a danger of bumping into things or even falling over.

The solution to the problem quickly became clear – lavatory landing lights! These were to be produced in softly glowing colours – rather like the directional lights on the floor of an aircraft that lead you to the emergency exits. Pink, luminous green and a soft chalky blue were used for the prototypes.

There were no major technical difficulties in producing the lights except that anything electrical near water has to be treated with great care. The answer was to run the thing from a low-voltage battery. The light was fitted inside the loo just under the rim and also round the lid. The makers claimed, with some justification, that the landing light didn't just make it easier for elderly people getting up at night but would also aid toilet training for children because it would make going to the loo fun. Plus, it would help them remember to put the lid down afterwards – leaving the lid up on some versions of the device created one light while putting it down created another. Blue for down and pink for up? – the customer was able to choose whichever colour scheme worked for him or her.

In the early 1960s a Japanese inventor came up with an even more bizarre device – the singing loo. Using a simple electric switch fitted to the loo seat he created a lavatory that would sing. As soon as the seat was lifted, the singing began and it didn't stop until the loo lid was replaced. Wagner was useful for persuading children always to remember to close the loo but for the constipated and others forced to spend unusually lengthy periods on the loo a ragbag of musical options was available, from Mozart to Kurt Cobain.

Of course, trying to create entertaining lavatories isn't entirely a modern thing – Thomas Crapper, who really did invent the first really efficient flushing lavatory, insisted that all his lavatories should be decorated with a tiny bee down at the bottom of the U-bend below the water. The reason was that the Latin for a bee is 'apis' (a piss). Clearly our Victorian ancestors were easily amused!

An advertisement in a British satirical magazine took this idea a stage further by offering to make lavatories in a range of custom-built designs – you could have the face of your worst enemy (or your best friend) painted on the inside of your loo. It is said that among the more popular options were Mrs Thatcher and Ronald Reagan.

PEACEFUL NIGHT

AMERICA, 1993

No one has ever disputed the fact that we are all better off if we get a good night's sleep every night. Eight hours is the recommended amount if you want to be healthy, wealthy and wise, but anxiety, stress and overwork – all virtually unavoidable in the modern world – mean that for many of us a really good night's sleep is a rarity. The biggest problem associated with sleep, of course, is bad dreams, which include everything from anxiety dreams to full-scale recurring nightmares.

Apart from not eating cheese and chocolate (popularly supposed to increase the likelihood of bad dreams), there's little that can be done but various inventors have become convinced over the years that they've come up with a solution. One American inventor constructed a nightmare catcher – a glass box containing just the right amount of chalk dust, sand and purified water to trap the vapours coming in through the window and afflicting the poor sleeper.

The only person who was sure the device worked was the inventor but it might well have been one of those things that would have worked if only enough people had believed in it – rather as people believe that peach nut oil is good for the hair. A slightly more scientific device was created by an American inventor in 1993. He was convinced that the biggest cause of nightmares and anxiety dreams was the pressure exerted by the mattress on the heart of the person sleeping. He therefore designed and built a mattress that would keep the sleeper in a position in which the heart rested continually above a hollowed-out area; in other words the heart would always be above a void in the mattress. To achieve this, the mattress had to be made so that the sleeping surface sloped slightly inwards from both sides. Not so much that the sleeper was made uncomfortable by being forced to roll down a hill, as it were, but just enough to cause a general tendency to sleep in the right position above the heart hollow. No one knows how many of these nightmare-prevention mattresses were built, but from any sane point of view it has to be agreed that it is amazing that any were built given one simple fact – there is not a shred of evidence that lying face down with the heart close into a mattress causes nightmares. The inventor was simply responding to a personal conviction based on wild fancy.

PET FRIENDLY

AMERICA, 1993

Americans must be among the most inventive people on earth and when you combine the rich fertility of their scientific imaginations with their famed passion for pets, you come up with some of the most wonderfully odd inventions ever heard of. The American patent office still has a carefully designed file for a device to hold the ears of floppy dogs at right angles to their heads – to keep their ears out of their food – and there's a special contraption that turns a dog into a sort of glorified shoulder bag. It's a dog coat that attaches right round the body of the dog from head to tail and is fitted with side pockets, a lead and a shoulder-carrying strap. When the dog gets tired you stop using the lead, put the shoulder strap over your shoulder and the dog becomes a bag. Then there are water-skis for pets – these have been designed for everything from pet squirrels to St Bernards – and even a diving suit for man's best friend. This again fits the dog like a coat but built into the back of the coat are two small oxygen tanks linked by a hose to a glass bubble helmet that fits neatly over the dog's head and is attached in a watertight fashion to the rest of the diving coat.

Other fine American pet inventions include a small kennel that straps to the dog's owner – it's designed for little dogs that like sleeping in their own space but also need the reassurance of knowing where their owners are at all times.

Perhaps the ultimate modern pet invention is the automated pet petter. Various versions of this device have been patented.

The most sophisticated – and absurdly expensive – was designed to get round the problem that certain types of dog hate being left alone all day while their owners are out at work. Rather than simply choose a different breed, the scientific American community tried to come up with a solution in the form of a device that gives the dog human attention without the need for a human actually to be around at all.

One patent filed in 1993 involves a pressure-sensitive area of flooring. This is fitted in a suitable corner of the dog's favourite room. Rising from the edge of the pressure-sensitive pad is a mast at the top of which an arm extends out over the pressure pad. The basic idea is that the first time the dog happens to tread or sit on the pressure pad, the arm descends and the dog gets a friendly pat. In true Pavlovian fashion the animal then quickly learns that every time it goes to that part of the room it will be rewarded with a pat.

The really sophisticated petter has a voice recording system that simultaneously says 'Good boy' or 'Well Done, Rover', or whatever the owner chooses to record. To make the petter really effective the hand on the end of the arm is fashioned from a remarkably lifelike latex which is impregnated with the scent of the dog's owner. The fingers are flexible and the advanced jointing mechanism of the arm means that as soon as it reaches any part of the dog's anatomy it moves the hand gently to right and left.

For dogs that hate being left alone – including many of the most fashionable breeds – the idea was a good one, but during trials of a simpler version the dog tended to bite the hand or jump up and grip the arm with its teeth every time it got bored. The more sophisticated versions were also incredibly expensive, but there's nothing to stop a very rich, very concerned pet owner commissioning his or her own pet petter: the technology is readily available – at a price.

IT'S A WIND-UP

ENGLAND, 1994

All across Africa and other poorer regions, staying in touch with the world is one of the biggest problems. Everything is fine while villages and small isolated communities have enough to eat but when famine strikes – or war or some other disaster – it's always the ordinary people who suffer most. As the African proverb puts it: when two elephants fight it is the grass that is crushed. Salvation for remote communities can come if they can be contacted – it might be that they need to know where the nearest feeding centre is, or where they need to move in order to avoid genocidal militias. But of course in remote areas among the world's poorest people there is no money to buy expensive electronic equipment such as laptops with internet access or mobile phones. And even where simple radios are just about affordable for the poorest, there is the huge difficulty of having the money to buy batteries and then finding them.

A British inventor decided that there had to be a solution to this problem and he came up with an idea that was initially rejected by every manufacturer he approached. The idea – which was to become hugely successful – was simple: why not build a clockwork radio? If the tension of a spring can provide energy to drive a watch or clock it should be possible to harness that same energy in a slightly different way and transfer it into a form that would drive a radio receiver.

Trevor Bayliss was the inventor in question. Even if it was possible to build such a strange device, what was the point, given that everyone who matters from a commercial manufacturer's point of view has easy access to electricity? Producers were inevitably less interested in an invention aimed at people who were among the world's poorest. Why should they fund the expensive process of building and refining a product for users who might not be able to afford it anyway? Bayliss was, however, determined not to give up – he had heard that the huge AIDS problem in southern Africa was exacerbated by a lack of information among poorer people whose lives might be saved if they could only be contacted regularly with information about the killer disease. Eventually the South African Liberty Life Group helped turn a brilliant idea into a reality.

The original large version is now owned by thousands of poor Africans – the very people it was aimed at in the first place – but ironically it has also become popular in a transparent plastic version in the West.

The radio is operated by cranking a handle round until enough latent energy has been stored in the clockwork mechanism to run the radio receiver and speaker. One hundred winds of the handle will give upwards of 30 to 40 minutes of air time and for the very poor, of course, there is the enormous bonus that they never have to visit a shop to buy batteries, nor worry about where the money is going to come from to pay for them.

FROZEN FINGERS

AMERICA, 1994

Bell's palsy – a condition that causes paralysis to one side of the face – can be brought on by driving in cold weather with your car windows open. If you do it for long enough – say several hours at a time – the muscles and nerves are so deadened that they effectively give up and may not start working again for weeks or even months. In rare cases the normal sense of feeling never returns and the victim's face has a permanent slack look as if he or she has had a stroke.

A similar problem arises if you happen to like driving with your arm resting on the car door and your hand draped stylishly out the window. Such a position can lead to damage to the nerves – again caused by the continual effects of very cold air passing over the skin – and complete loss of muscle control.

An inventor who was convinced that there were large numbers of drivers who really couldn't be persuaded to drive with their windows shut whatever the conditions, decided to find a solution that would enable cool drivers to carry on driving in style whatever the conditions.

He came up with the peculiar-looking arm mitten – a cumbersome padded sleeve that you donned on starting your journey just for the pleasure of being able to dangle your arm out of the window. The sleeve looks like a giant oven glove that reaches three quarters of the way up your arm. Though it undoubtedly does its job it has the slight drawback of looking completely and utterly ridiculous (which meant arm dangling no longer appeared or felt cool) and also makes steering a little awkward – you either whip your arm out of the sleeve when you need both hands on the steering wheel or try to steer everywhere with one hand.

HIGH FLYERS

AMERICA, 1994

Early aviators simply would not have believed the sophistication of later aircraft. It's not just that you can now fly in great comfort from Europe to Australia in a day, nor that aircraft can carry hundreds of passengers at a time over those huge distances. The real advances in the science of aircraft engineering are to be found in military aircraft. Where scientific energy has traditionally concentrated on creating aircraft that remain stable whatever the conditions, military aircraft designers have in recent years deliberately tried to introduce inherent instability.

This may sound completely mad but it actually represents the saner end of aircraft invention. The idea behind the usefulness of instability for flight comes from bird observation. Birds are supremely manoeuvrable in flight because their on-board computers – their brains – allow continual and very rapid adjustments to wind speed and direction. When the bird needs to move quickly – to avoid a predator or to catch prey – the brain allows the normal flight pattern to be disrupted and the bird goes into free fall or executes a dramatic turn. Jumbo jets and other commercial aeroplanes are the complete opposite – they are so stable that they respond relatively slowly to the pilot's controls. That's fine for a plane that wants to cross the Atlantic in comfort but military planes are better off in combat if they can go into sudden free fall to avoid an enemy or jink and twist second by second to outmanoeuvre a pursuer. Modern computers make this possible because they can adjust the controls in microseconds to keep the plane on an even keel during normal flight while allowing the option of sudden chaos when rapid manoeuvres are needed.

The use of computers for this sort of thing has led to at least one bizarre invention, however, that is applicable to passenger aircraft – the paddle-driven aeroplane. At the moment, lift for all aircraft except helicopters is provided by large wings. An alternative is to have lots of short wings arranged in the pattern we see on old-fashioned paddle steamers. The mass of wings mean there's plenty of theoretical lift but until modern computers it would have been impossible to translate the lift from these spinning blades into an overall lift stable enough to keep a large aeroplane in flight. Of course, such a plane has yet to be tested but the patents are there if anyone is feeling really brave and computers can, at least in theory, cope with the huge complications of the physics involved.

CONDOMS AT CHRISTMAS

AMERICA, 1994

Sex has only recently become an area of major interest for scientists and inventors. The problem in the past, particularly in Britain, was that any discussion at all of sex was completely taboo – which led to some very amusing situations. Parliament, for example, refused to outlaw lesbianism because their Lordships were convinced that no decent Englishwoman would ever have heard of such a thing anyway and that if they passed legislation outlawing it the net effect would be to publicise lesbianism and thereby tempt previously innocent matrons and their daughters.

But once discussion of sex became possible, the floodgates were open and a rich source of opportunity opened up to the inventor. Apart from relatively innocent devices – like the glove with two sets of fingers but a shared palm (to enable lovers to walk hand in hand in cold weather) – there was a coition chair, basically a wooden chair fitted with a back support on either side. The instruction manual advises the couple intending to use the chair to sit facing each other with their backs supported by their respective chairbacks. The seat itself is so narrow that, undressed and facing each other, the couple would automatically be in just the right position. The chair came in a vibrating version for the really enthusiastic!

There have been numerous black boxes designed to send electrical charges through the genitals, including a particularly fine example invented by the artist Keith Vaughan (1912–1977), which allowed the current to be adjusted to suit all moods.

The ultimate mad sexual invention, however, has to be the re-usable advent calendar. Each of the 25 doors leading up to Christmas conceals a different coloured and or flavoured condom to be used to liven up those long, cold December nights. On Christmas Day itself the advent calendar door reveals a bright red condom with a white band round the middle and a white bobble on top – yes, you've guessed it, a Santa condom.

INVISIBLE PET

AMERICA, 1997

We all love our pets but some people clearly love them more than others if the effort to make their lives happier or longer is anything to go by. All over America (not to mention Europe) it seems dogs are catered for with poodle parlours, slimming centres, canine nail manicure shops and even doggy psychotherapists. Cats in Britain often get similar treatment and every year there are cases extensively reported in the newspapers of elderly cat-owners leaving millions of pounds to their pets. Today you can even have your pet cryogenically frozen in the hope that some future technology will discover how to bring Fido back to life complete with the character, the tail wagging and the memories he had in life. The passion for pets, though seen by some nationalities as decadent, is from other points of view touching in the extreme.

Scientists have also made their way to the pet emporium with nappies for everything from cats to canaries, little wheels for dogs that have lost their back legs, coats and harnesses of various kinds and a vast array of cosmetics, perfumes and pet-specific medicines.

But the concept of pet culture was taken one final bizarre step by the American inventor Daniel Klees in 1997. The imaginary pet leash is a long, stiff, lightweight tube moulded to look like a real leash. On the end of the leash is a stiffened dog harness (available in various sizes according to whether the customer requires an imaginary Great Dane or an imaginary Chihuahua).

The imaginary pet leash gives the impression that a dog is being taken for a walk but where the dog ought to be, there is in fact nothing at all. Mr Klees' inventiveness doesn't end there – to add to the verisimilitude of the whole thing he even added a pre-recorded dog noise tape which can be operated by a cunningly concealed button on the handle of the leash. The speaker for the tape is attached to the dog collar and the tape can produce howls, barks, yelps of delight, snarling noises and whimpering. And all at the touch of a button.

Of course the real advantage of the imaginary pet leash is that the owner has a reasonably pet-based reason to go for a walk but suffers none of the drawbacks of real pet ownership – huge vet bills, chewed furniture, terrified local cats and the horrors of regular poop scooping.

RUDE AWAKENING

JAPAN, 1998

Falling asleep when you really don't want to is a major problem, particularly in the West where long hours and stressful office jobs make tiredness a permanent condition for many. This manifests itself most disastrously in the tendency for tired office workers and lorry drivers to fall asleep at the wheel. A huge percentage of fatal road accidents is estimated to be caused by this phenomenon and anyone who thinks the problem can be solved by opening the car windows, shouting, singing and listening to the radio at full volume needs to think again because the condition is highly specialised and not directly related to normal sleep patterns. In fact, falling asleep at the wheel is a unique phenomenon about which comparatively little is known; it is almost as if the victim has been drugged. Large amounts of caffeine will work to prevent the problem providing the hypnotic routine of driving (particularly on motorways) is broken first by a stop, but many people dislike taking large amounts of caffeine.

One answer comes from a Japanese company. The Thinky Corporation of Tokyo decided that a solution to the general problem of sleepiness could be achived without the need for drugs. After several years of research into what really gets the brain going and in such a way that sleep becomes the very last thing possible, they came up with the splendidly named Brain Buzzer.

This looks rather like a large electric toothbrush but whereas on a toothbrush the narrow end tapers to a brush head, on the Brain Buzzer the narrow end simply tapers. The narrow end is in fact a bitable rubber wand. According to the instructions – meticulous and thorough as on all Japanese products – the user should grasp the thick, handle end of the buzzer firmly before biting hard on the narrower end. With the teeth safely clamped into position, the gripping hand flicks a switch and the narrow wand begins to vibrate violently.

The effect on the brain is apparently to knock out all the cobwebs and of course all thought of sleep. The only drawback to this highly idiosyncratic device is that unless the user keeps his or her mouth tightly closed while using it, the noise of the vibrations echoing in the mouth is very loud. It also sounds similar to an unmentionable bodily function. Hence the instruction on the box: to avoid annoying fellow passengers, co-workers or others, always keep the mouth tightly closed during operation.

KISS AND TELL

AMERICA, 1998

Everybody loves kissing someone attractive. We've been doing it for countless centuries and we're not likely to stop, but increasing knowledge about how diseases are passed on (often by droplet infection from nose or mouth) have made some poor souls rather squeamish about kissing at all without some sort of protection.

Certain cultures are also rather more sensitive to these things than others. Take the Japanese, for example. It is probably fair to say that the Japanese are almost obsessed with their health – antibiotics are available over the counter in Japan and many Japanese take them routinely every day rather as one might take a vitamin pill. The net result is that antibiotics are daily becoming less effective and may eventually become completely useless. Taking them every day and for trivial infections simply trains the bugs to be resistant to them.

At least one inventor, concerned at the whole infection issue and the risk of making one of the world's most effective medicines redundant, came up with a solution: the kissing shield.

The shield looked rather like an old-fashioned fan with a handle attached to a wide curving mesh stretched between struts. The very fine mesh allowed the kisser to see the about-to-be-kissed (and vice versa) and the pressure of the kiss would certainly be felt through the mesh. The neat trick, however, was to impregnate the mesh with a strong odourless disinfectant so that any germs that tried to get through would die in the attempt.

For some inexplicable reason the kissing shield never took off – perhaps we are all so addicted to kissing that we are prepared to take any risk for a moment's pleasure.

TONE WHILE YOU WALK

AMERICA, 1998

The pressure on everyone to look good which – in the West at least – means looking fit and thin, has been a goldmine of opportunity for inventors. From devices that guarantee to strip away the pounds while you sleep to contraptions that use the power of the spirits to help you shape up, someone somewhere is usually working on an idea to help us get just that little bit closer to physical perfection.

Among the funniest and perhaps on the face of it most plausible is the fitness and toning suit designed in America to make everything you do in the course of moving about a source of strenuous exercise.

The patent is based on a tight-fitting one-piece suit made in a thick, breathable material that leaves only the hands, feet and head free. Running in all directions over the suit, back and front, are what look like thick cords of material, about one inch in diameter. These ribs are actually an intricate criss-crossing system of carefully calibrated devices for making the suit resist movement in any direction. When you walk, all the ribs act together to make walking difficult not because walking is inherently difficult but because any movement you undertake in the suit meets resistance. Acting against that resistance uses energy, thereby helping you lose weight and shape up.

Like so many zany inventions it sounds great in theory and the idea was meant to appeal to busy executives on the basis that they could get the same toning and weight-loss effect from a twenty-minute resistance-suit walk as they could from a normal two-hour walk.

In practice, of course, the suit simply didn't seem to make enough difference to justify the cost of buying it in the first place or suffering the stares of passers-by. The designers clearly hadn't even considered this kind of adverse reaction: their literature mentions the fact that the suit is so beautifully designed and colourful that it could easily be worn all day every day to the office, the shops and while visiting friends. The designers also point out that the suit is perfect for formal evening wear. Even by the standards of advertising copy, that is probably going a bit too far!

A similar exercise suit made in Britain was tested out by a team of student volunteers who complained that the tension on the cords was impossible to adjust precisely – they either felt completely trussed up and unable to move at all or movement was too unrestricted and any exercise benefit thereby lost. The suit also made several students faint from overheating and loss of body fluids.

Back in America the wealthy found an alternative to the exercise suit that has pretty much the same effect – serious runners in Central Park in New York and elsewhere in fitness-crazy parts of the US can regularly be seen running about ten feet ahead of their personal trainers. The trainer holds a set of reins that are attached to a harness on the back of the wealthy runner. The runner is constantly pulled back by the trainer and has to lean into the harness and exert twice the normal effort to propel both himself and the trainer forward. Sounds mad, but it works!

POWER TONGUE

AMERICA, 1999

The Italians almost certainly invented ice cream but the Americans, rarely happy to leave things as they find them, now make more varieties of ice cream than any other nation on earth. Indeed one or two famous American ice cream brands – brands that are stuffed with the absolute maximum of cream and sugar – have been blamed for the fact that a vast proportion of Americans are gigantically fat.

And then there is the whole science of eating ice cream – cones, wafers, cups and sticks must have seemed as dull as ditchwater to the American inventor who decided to create the motorised ice cream cone holder.

This was thought up towards the end of the 1990s as a novelty to end all novelties. An elaborate battery-powered mechanism was fitted in the base of an oversized plastic cone shaped and coloured to look like a real, if over-large, cone. The mechanism when activated caused the top half of the cone – the part in which a real ice cream cone was designed to sit – to spin rapidly. The idea was to eliminate all that tiresome and exhausting turning by hand. The motorised ice cream cone holder made keeping your ice cream smooth and even round the edges very easy indeed. But the inventor of the device had even higher hopes for it. In his patent he said he hoped that children would be encouraged to carve interesting shapes in their ice creams now that the dull mechanical business of turning them had been eliminated.

HURRICANE HOUSE

AMERICA, 1999

Every summer countries across the Caribbean from Cuba, Haiti and Trinidad to the American mainland live in dread of the next hurricane. Summer is the hurricane season in this part of the world and very little can be done to alleviate the effects of winds that can reach two hundred miles an hour. The predictability of atrocious weather can be judged by the fact that citizens in the line of the summer hurricanes cannot get buildings insurance – who, after all, would insure a house known to be in the path of something as potentially destructive as a hurricane?

The solution has long been seen as better buildings, but in poor countries expensive building materials are simply out of the question, which may well be why one US scientist came up with one of the strangest ever ideas for a hurricane-proof house. The patent was filed in 1999 and though it has not yet been tried (for reasons that will become obvious), there is no doubt that it would alleviate many of the problems associated with life in the hurricane belt.

The patented house is actually an old jumbo jet (or other large airliner) fitted to a circular concrete tower. The interior of the plane is kitted out with partitions and rooms just like a normal house and these rooms continue in the circular tower on which the plane rests.

The idea behind what seems like a really mad concept is that aeroplanes are specifically designed to survive hurricanes;

their shape, when turned into the wind, offers the least resistance to the worst the elements can hurl at them.

The aeroplane house is actually designed to work like a weathervane. You can turn it in any direction you like when the weather is good, but when a hurricane blows you leave it on autopilot, as it were, and it turns to face in the direction from which the worst of the weather is coming. Airliners really can and do survive hurricanes so the airliner house would almost certainly work. But, as with every invention, there are difficulties. First, airliners are expensive: even very old planes cost a great deal more than the average Caribbean family could probably afford. Then there is the cost of building the concrete tower and the turning mechanism and finally, what on earth would the planning authorities in hurricane-affected countries make of the idea of whole villages made up of aeroplanes on towers?

DAD SADDLE

AMERICA, 2002

The first person who gave a child a shoulder ride has a lot to answer for. The ache caused by twenty minutes of heavyweight four-year-old plonked on the sensitive neck muscles has to be experienced to be believed, but few parents can resist the pleas of their offspring. Seeing a gap in the market, a number of inventors have tried their hand at making the burden of parenthood that bit easier to bear.

We've all seen rucksacks with baby's head poking out the top and there are all kinds of slings, pouches and papooses, but at the more extreme end of baby carrying there are some wildly eccentric contraptions. Take the Daddy Saddle. Invented in America, this actually does a pretty good job in at least redistributing the weight of the child to cause least damage to the parental back – but the price you have to pay for extra comfort is undoubtedly extra embarrassment because the saddle really does look like a saddle. Let's assume you don't mind looking like a pantomime horse, and find out how it works.

Made of leather and carefully designed to sit snugly on the hips, the Daddy Saddle is fitted with a pair of stirrups attached to the saddle by leather straps. The stirrups – and they resemble the real thing – are placed either side of the saddle so that when the child decides she wants a ride she simply climbs aboard and stands with her feet in the stirrups and her arms round the parental neck.

In theory this is a big improvement on the straightforward unassisted shoulder carry or piggyback because the weight pushes straight down on the hips rather than pressing on sensitive nerves and muscles in the neck.

From the almost plausible we move quickly to the totally implausible. In the same year that the Daddy Saddle was created, another inventor came up with the Daddy Car Bucket seat. This is a plastic seat fitted with a steering wheel and windscreen. It actually looks like a small plastic car attached to the top of a rucksack. The inventor planned to offer a range of similar products – a helicopter seat, aeroplane seat and even a horse with saddle.

Elaborate straps and struts were needed to produce a strong enough rucksack base to hold the car or aeroplane above the parent's shoulders and head, but a mini safety belt was included so that the child couldn't fall out of the seat. Once in position, the child would find he was riding high above his parent, surrounded by an authentic-looking cockpit complete with steering wheel. What could be more fun? Of course for the parent it would be even worse than a straightforward shoulder ride, if for no other reason than it would be even more fun for the child, who would be very reluctant to be put down at all.

ANGRY MOUTHFULS

AMERICA, 2002

The biggest health problem of modern times in the West is probably the problem of obesity. While half the world barely gets enough to eat, people in the developed world don't seem to know when they've had enough. Obesity is defined not as being overweight but as being so seriously overweight that there are major health implications.

Obesity in children seems to be an even greater cause for concern – blamed variously on computer games, the internet and a passion for TV. Combine a passion for sedentary interests with an industrial food production system that has made fat-rich and sugar-rich foods incredibly cheap and you have a huge problem. Among the cures for the more seriously overweight are having your jaw wired up (that way you just can't get the food in even if you want to) or having part of your intestine removed so that however much you eat only a small percentage of your calorific intake will be absorbed and end up as body fat.

But these are drastic measures which themselves have potentially significant health implications. A better solution has long been sought by both scientists and doctors. Health education programmes do work for many but for those who don't want surgery but would still like to lose weight there is at least one apparently sensible but actually quite dotty invention: the fork that tells you off if you eat too much.

Invented in America in the late 1990s the fork is actually a miniature traffic light. It looks like a conventional fork but has a set of lights on the handle facing the user. One light is green, the other red. Inside the hollowed-out handle is a complex piece of circuitry that enables the fork to 'know' when the user has either had too many mouthfuls or is eating too fast.

Eat carefully and at a moderate pace and the fork's green light stays on, but if you start to gobble your food the red light comes on and the fork emits a warning beep until you slow down again. Pick up too many forkfuls and the beep gets louder and the warning light begins to flash. The fork can be pre-programmed to allow a certain number of mouthfuls and no more.

In trials the fork worked well but serious eaters found it was such a bully that it made them angry and they either ignored it and deliberately ate more to cause it to flash furiously at them or they simply threw it away and used a conventional fork.

DRIVEN TO DISTRACTION

AMERICA, 2003

Mad inventions are not just a product of Victorian cranks and steam train obsessives – in fact if anything today's inventors are even madder than their ancestors. Which explains why the number of crazy inventions increases every year. In America the patent office is permanently awash with electric toothpicks, singing sunglasses and bicycles that massage your bottom while you pedal. And every year brings one or two inventions that seem insane but lead to something really useful and wonderful. The clockwork radio invented by Trevor Bayliss is a case in point.

All inventors are drawn towards inventions that make life easier – the theory being that if you create something that saves time, energy and money (or any combination of those three), your invention will make you, the inventor, a great deal of money. Which brings us to the motorised picnic table. Invented in America as recently as 2003, the picnic table is fairly straightforward-looking but conceals a motor and a set of all-terrain tracks or wheels. With a Victorian-style tablecloth (very long to hide unseemly legs) the motorised picnic table would be indistinguishable from any ordinary table except for the fact that it had four seats permanently attached and a small steering wheel. If you happen to have a big garden or a few fields the family could sit round the table for a picnic lunch while slowly touring the estate. Those without a garden or several hundred acres could simply load the table on to a conventional vehicle and then carry it off to the countryside. Once you'd found your beauty spot and laid the table for lunch you could start eating and driving round at the same time.

The motorised picnic table did have its devotees but it tended to cause motion sickness in children and some adults and the whole thing was prone to disappearing down potholes, off cliff edges or into rabbit holes if the driver was too busy with his own fish and chips.

A REAL STUD

AMERICA, 2003

Almost everyone hates wearing glasses. People who don't wear glasses think this has more to do with vanity than anything but in fact the real reason spectacle-wearers get annoyed is that having an ounce or more of glass and metal perched on your nose for hours on end can be uncomfortable.

Even when you get used to the idea that you have to wear them, there's the sheer awkwardness of trying to play sport with them on or wearing them in bad weather. And then there are the arms – the bits that hook over your ears. It's been estimated that only about one in ten pairs of spectacles has arms that fit the wearer. Ideally the arms should be just short enough to stop the glasses slipping down your nose but just long enough to avoid giving you the sensation that someone is trying to pull your ears off. It's a balance that's hard to achieve and at least one American inventor convinced himself that he had the answer: eliminate the arms altogether and replace them with studs.

The idea had the advantage of linking with the hugely popular fashion for body piercing because the studs that the glasses were designed to attach to had to be gunned into position through the eyebrows. If you were going to have studs anyway, why not have the kind that would kill two birds with one stone and thus eliminate the need for those irritating spectacle arms and put yourself at the forefront of fashion? The design for the stud glasses was careful and meticulous.

Frames were lightweight with ingenious snap on and snap off magnetic pads. Early versions used a too-powerful magnet, which meant the wearer got into something of a tug of war with his or her own head before the glasses could be removed. Happily, by the time the glasses reached the marketplace these teething problems had been sorted and there was – surprisingly enough – a market for them, but it was a tiny market. The problem was that the bulk of the market for glasses is and always has been among the middle aged and elderly, who were always somewhat unlikely to rush out and get themselves fitted with eyebrow studs. As a result another great idea died in its infancy. But along California's sunshine coast there are still to be seen pairs of stud-fixed spectacles, only now they are like rare birds and when their owners finally die the stud-fixed spectacle will disappear for ever.

THE PERFECT SLEEPING PARTNER

JAPAN, 2004

Women all over the world are said to complain that, much as they may love their husbands, they find sleeping with them mostly rather annoying – husbands fidget, snore, roll back and forth, become amorous at odd and unpredictable times through the night, and generally make a good night's sleep difficult or impossible.

A Japanese inventor has come up with a solution that is proving hugely popular, at least in the home market: it's a pillow shaped like a man and with an arm fitted at just the right angle to provide a comforting cuddle.

One satisfied customer was quoted by the makers as saying: 'It's much better than the real thing – warm, comforting, never argues or fidgets. I've slept better since I got it than I have since I was a child.'